Why Kenyan Coffee?

Sam Mungai

Melbourne, Australia

Mungai C/- Intertype Publish and Print
Unit 45, 125 Highbury Road
BURWOOD VIC 3125
www.intertype.com.au

Ordering Information:
Quantity sales. Special discounts are available on quantity purchases by corporations, associations, and others. For details, contact the "Special Sales Department" at the address above.

Why Kenyan Coffee?/ Sam Mungai. —1st ed.
ISBN 978-0-6456620-5-4

Contents

Forward

This book has been written with a view to tracing and preserving the history and evolution of the Kenyan coffee industry. It focuses on the uniqueness of the Kenyan coffee production, processing, and marketing aspects. The book borrows from factual, historical, administrative, technical, marketing and experience aspects of the industry. It also seeks to enlighten the new generation of farmers on the evolution of coffee processing and marketing that can assist in understanding and tackling challenges facing the industry.

The book also endeavors to contribute knowledge of the Kenyan coffee industry to the coffee world

CHAPTER 1

INTRODUCTION

History of Coffee

History alleges that coffee was first discovered in 850 A.D at a place called Kaffa, in the southwestern part of Ethiopia and later spread to Yemen. It is believed Ethiopia was the origin of the word bean, derived from the word 'bun' in the Ethiopian language while the word coffee is said to have been derived from the word Kaffa. It is alleged contrary to how coffee is roasted, ground and drunk as a beverage today, in the early days of coffee history, coffee beans were chewed as a stimulant. Coffee brewing is said to have taken root spreading around the 13th century. History alleges the first coffee shop was opened in Constantinople in the middle of the sixteenth century followed by another one in London a century later. Heavy taxes were imposed on coffee shops without success to curtail their spread as they were viewed as politically seditious.

Mokha (Mocha) a port city on the red sea in Yemen was then at the heart of the coffee trade.

Coffee houses became important centres for exchange of information. According to an Article by Emmy Blotnick December 14, 2013, there were five attempts to ban coffee in Mecca in 1511, Italy in the 16th century, Constantinople in 1623, Sweden in 1746 and Prussia in 1777. The bans were instilled by acts like confiscation of cups and dishes, while in Constantinople it was by beating for the first offence, while the second offence resulted in offenders sewn into a leather bag and thrown into waters of the Bosporus. Despite all these attempts coffee spread globally becoming the second most traded commodity after oil.

Today coffee is facing other kinds of threats including price volatility, and climate change.

Chapter 2

KENYA AS A COUNTRY

Kenya acquired its name from the snow-capped Mount Kenya lying south of the Equator known then to the local inhabitants (the Kikuyu) as 'Kirinyaga'. The early European travellers could not pronounce the name Kirinyaga properly, shortening it to 'Kenya'. Since then, the mountain became known as Mount Kenya. Subsequently, the country's name became Kenya. After the scramble for Africa by the European colonial powers, Kenya was declared a British protectorate in 1902 against a strong resistance by Africans.

With the expansion of European settlement in the highlands, Kenya was finally turned into a British colony. The confiscation of land for white settlement by the colonial administration led to an uprising by the African resistant movement namely the Mau Mau which is credited for Kenya gaining Independence.

Mount Kenya

Photo courtesy of Boniface / Unsplash

Kenya lies within the Equator, with Sudan and Ethiopia to the North, Uganda to the West, Somalia to the East and Tanzania to the southern part of Kenya.

The country has an area of 580,876.3 square kilometres with an average population of approximately 47,564,296 (Kenya National Bureau of statistics 2019).

Kenya is predominantly an agricultural country and experiences an Equatorial climate which in the past presented weather conditions with a well-defined rainfall pattern. This has changed due to climate change occasionally presenting unpredictable and inconsistent distribution of rainfall.

Coffee in Kenya is grown on volcanic soil. These are some of the factors that have contributed to the good quality coffee that Kenya has produced over the years. Kenya coffee growing regions also enjoy cooler months due to high altitude.

Production has declined over the years oscillating between 40,000 to 50,000 metric tonnes per year. There has also been reduction in quantity of high-quality coffee.

CHAPTER 3

HISTORICAL BACKGROUND OF COFFEE IN KENYA

Photo courtesy of Nairobi Coffee exchange

Kenya coffee is said to have originated from Arabica varieties grown in Yemen, namely Mocha Arabica. These seeds were introduced to East Africa in the 1800s.A British East India Company bought some seedlings in Yemen and sold them to the French Missionaries. The seedlings were eventually planted in the African Mainland resulting in a coffee with a unique flavour characteristic which became the basis of the East African Arabica coffee varieties.

During these times the Scottish missionaries were experimenting with Mocha at their station in Kenya while the French were working in the present-day Tanzania.

According to the original journal of Saint Austin's mission in Nairobi, Brother Solanus Zipper introduced seeds to the Nairobi mission from Morogoro in 1897.

In 1900 Saint Austin mission had its first flowering crop. The seeds introduced were thought to be Mocha but were no doubt a hybrid with some bourbon coffee characteristics. (Kahawa -Kenya Gold 2005 by Jeremy Block .et)

By 1914 the mission had 52,000 mature trees and could not meet the demand of seedlings from the settlers (white farmers).

One source claims coffee was first planted in Kenya at Bura Taita hills in 1893 and at Kibwezi under irrigation in 1900. Around 1904, coffee was introduced to Kikuyu near Nairobi leading to the introduction of commercial coffee farming.

In 1920, coffee farming was introduced into the Rift Valley, followed by Meru and Kisii in 1935. By the year 1946, coffee farming was only an activity of white farmers (known as settlers). In 1952 with the pressure for Independence mounting, extensive small scale growing of coffee by locals was initiated in all the suitable areas in the country. This was later followed by rapid expansion of land for growing coffee up to the time of independence in 1963 and post-independence.

This led to the emergence of two important sectors in the Kenyan Coffee industry, namely the Plantation sector comprising of large coffee estates (farms) and the Co-operatives sector comprising of small holders with small farms. By 1965, the Co-operative sector caught up with the Plantation sector and by 1967, was leading in coffee production controlling almost two thirds of the total acreage under coffee.

Co-operative societies were then organised under a body by the name, Kenya Planter's Co-operative Union (KPCU), which was formed in 1937 for the purpose of milling and marketing of coffee. At this time, the coffee industry was under the supervision of the Ministry of Agriculture but was later placed under the supervision of the Coffee Board of Kenya, an institution which was formed in 1933 by an Act of Parliament.

The functions of the Board included licensing, marketing, regulating, and promoting coffee, while the ministry of Agriculture formulated policies.

PRODUCTION AND AREA UNDER COFFEE OVER 57 YEARS

		ESTATE		SMALLHOLDER (Co-operative Societies)		COUNTRY TOTAL	
NO	**YEAR**	**Area (Hectare)**	**Production (Metric Tonnes)**	**Area (Hectare)**	**Production (Metric Tonnes)**	**Area (Hectare)**	**Production (Metric Tonnes)**
1	1962/63	35,538	24,800	49,121	16,000	81,659	41,400
2	1963/64	32,538	28,405	49,121	15,373	81,669	43,778
3	1964/65	32,423	22,393	51,577	14,774	84,000	37,167
4	1965/66	32,267	25,683	53,097	25,523	85,364	51,296
5	1966/67	31,964	25,231	52,450	27,558	84,414	52,789
6	1967/68	31,188	13,246	54,087	20,515	85,275	33,761
7	1968/69	30,690	22,342	54,984	23,264	85,674	45,606
8	1969/70	29,903	26,521	54,057	26,275	83,960	52,796
9	1970/71	29,900	28,600	53,800	26,302	83,700	54,902
10	1971/72	29,535	29,984	55,555	28,362	85,090	58,346
11	1972/73	29,535	39,956	55,287	34,734	84,222	74,690
12	1973/74	29,129	31,152	55,600	40,864	84,729	72,016
13	1974/75	28,603	29,985	57,786	35,464	86,389	65,449
14	1975/76	28,603	37,675	56,595	36,135	85,198	73,810
15	1976/77	27,821	49,685	56,600	47,660	84,421	97,345
16	1977/78	30,888	33,685	56,600	47,744	87,488	81,429
17	1978/79	29,195	26,809	62,574	46,079	91,769	72,888

		ESTATE		SMALLHOLDER (Co-operative Societies)		COUNTRY TOTAL	
NO	YEAR	Area (Hectare)	Production (Metric Tonnes)	Area (Hectare)	Production (Metric Tonnes)	Area (Hectare)	Production (Metric Tonnes)
18	1979/80	31,232	39,109	71,172	51,900	102,404	91,009
19	1980/81	32,861	34,744	84,710	64,007	117,571	98,751
20	1981/82	33,635	34,392	97,473	52,531	131,108	86,923
21	1982/83	33,605	32,981	100,967	52,469	134,572	85,450
22	1983/84	35,711	54,258	114,235	74,682	149,946	128,941
23	1984/85	35,711	28,922	116,328	67,717	152,039	96,639
24	1985/86	38,627	45,542	117,677	68,385	156,304	113,927
25	1986/87	38,487	36,881	116,056	67,907	154,534	104,288
26	1987/88	36,677	44,506	116,353	84,356	153,030	128,862
27	1988/89	39,579	38,649	116,087	78,340	155,666	116,989
28	1989/90	38,084	34,356	117,459	69,483	115,543	103,839
29	1990/91	38,024	37,520	120,238	41,977	158,262	79,497
30	1991/92	38,160	32,781	120,563	42,426	158,723	75,207
31	1992/93	38,372	33,037	122,660	39,747	161,032	71,787
32	1993/94	38,372	38,735	122,660	52,264	161,032	90,999
33	1994/95	38,372	40,109	122,660	56,885	162,410	96,994
34	1995/96	39,750	29,737	122,660	38,260	162,470	67,997
35	1996/97	39,750	22,050	122,720	33,584	167,398	55,634
36	1997/98	40,127	28,591	127,271	40,086	167,398	68,677
37	1998/99	40,127	38,585	127,271	62,265	167,398	100,850
38	1999/00	40,127	26,743	127,271	23,800	167,398	50,543
39	2000/01	42,000	26,900	128,000	25,000	170,000	51,900
40	2001/02	42,000	23,073	128,000	28,822	170,000	51,895
41	2002/03	42,000	21,417	128,000	34,026	170,000	55,443
42	2003/04	42,000	18,473	128,000	29,958	170,000	48,431
43	2004/05	42,000	20,745	128,000	24,500	170,000	45,245
44	2005/06	42,000	21,257	128,000	27,046	170,000	48,303
45	2006/07	42,000	25,000	120,720	28,368	162,720	53,368
46	2007/08	40,680	19,740	122,040	22,260	162,720	42,000
47	2008/09	24,521	26,650	84,263	29,370	108,784	54,020
48	2009/91	24,521	19,720	84,263	22,280	108,784	42,000
49	2010/11	24,521	16,690	84,263	19,630	108,784	36,629
50	2011/12	24,606	25,001	85,189	31,256	109,795	49,960
51	2012/13	24,606	13,773	85,189	26,092	109,795	39,865
52	2013/14	24,700	16,805	85,300	32,652	110,000	49,475
53	2014/15	26,067	14,808	87,433	27,230	113,500	42,038
54	2015/16	25,800	15,267	88,200	30,758	114,500	46,121
55	2016/17	25,900	14,151	88,800	24,468	114,700	38,620
56	2017/18	25,155	10,994	90,415	30,381	115,570	41,375
57	2018/19		14,103		30,881		44,984

The table shows the area under coffee cultivation and the production growth both in the Co-operative and Estate (plantation) sectors from 1962 just before independence to the year 2019.

Source: Agriculture and Food Authority (Coffee Directorate) coffee yearbook season 2018/19

The graph below shows coffee production in the major African growing countries has not changed significantly from the early 1990's to present day when compared to Brazil and Vietnam.

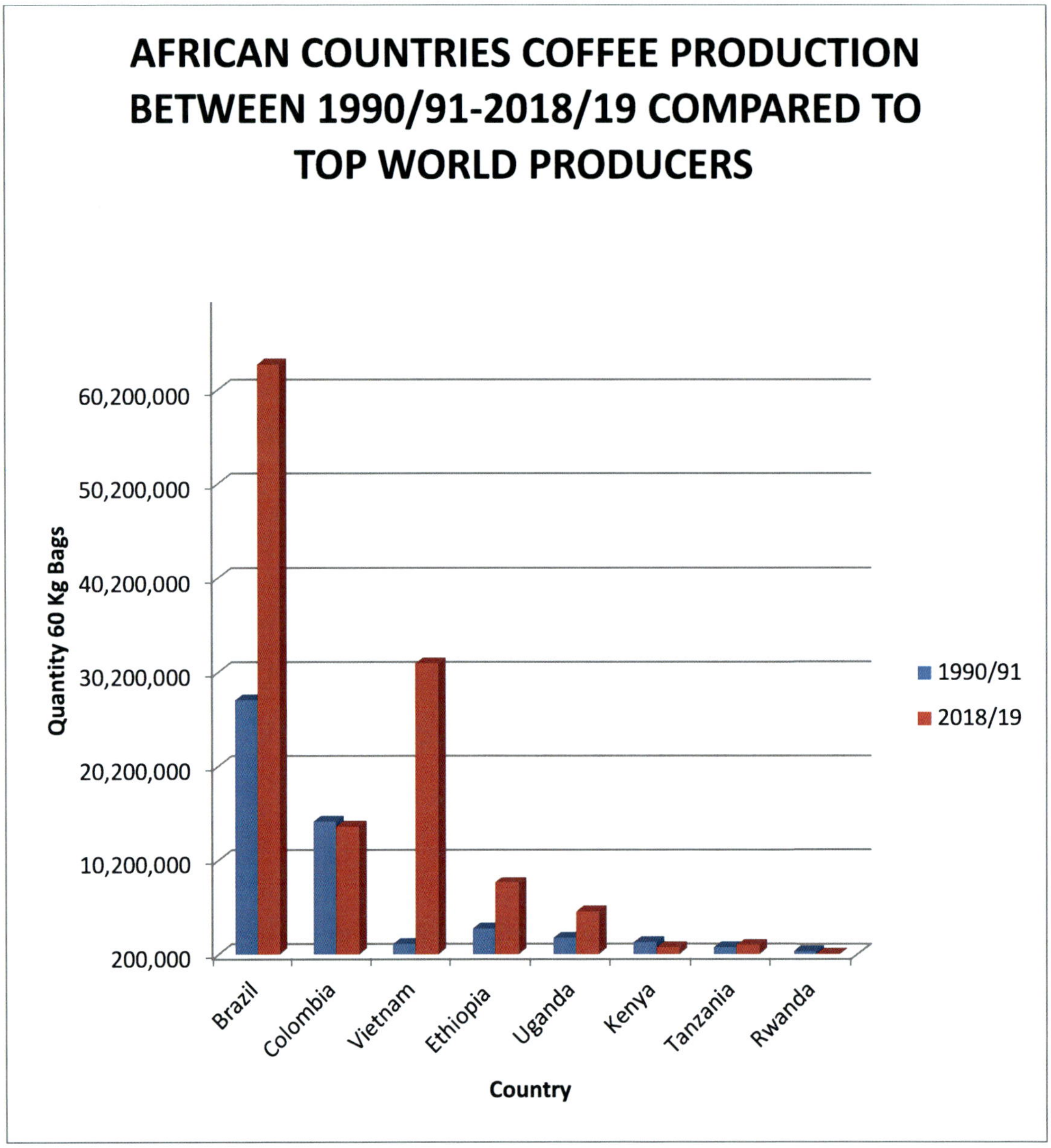

Source: ICO Different origins production growth and decline

CHAPTER 4

MAIN COFFEE GROWING COUNTIES IN KENYA

Coffee growing in Kenya was mapped into coffee districts with each district identified by a Reference Number. With Kenya adopting a devolved system of government the mapping is now based on counties which are the administrative units of county government. There are forty-seven counties in total. There is also a Mill identification number and an additional Outturn Number assigned at the Dry Mill to every consignment delivered, based on a first come basis. This Outturn Number appears on all documents for a consignment.

Part of the Outturn number includes a reference as to whether the coffee has come from an Estate or a Co-operative Society. Reference Numbers (County Number, Mill Number and Outturn Number) are used to trace the origin of coffee from a given County.

MAIN COFFEE GROWING COUNTIES IN KENYA

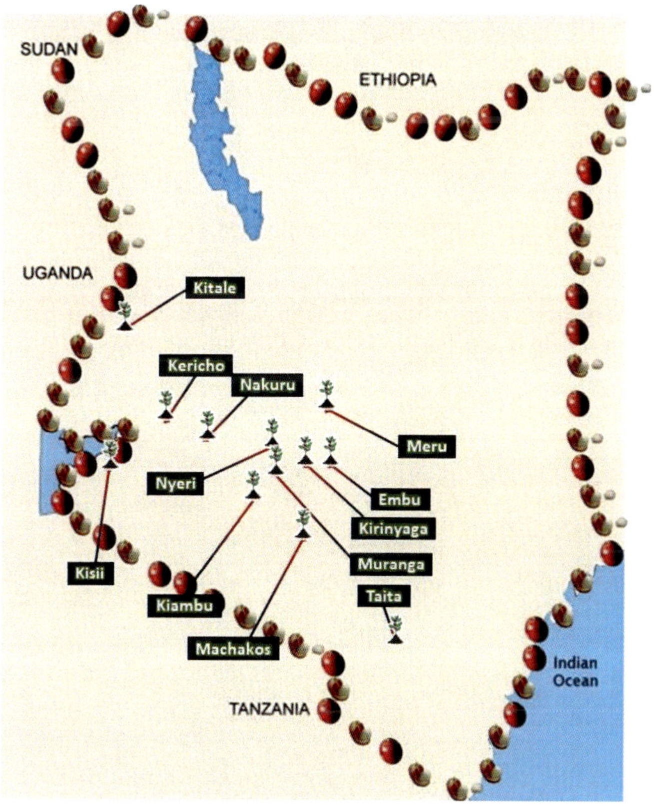

Kenyan Main Growing Coffee Counties

Central Kenya Region

Central Kenya is the main coffee growing region in Kenya producing about 80 % of the total production. It comprises Kiambu, Nyeri, Kirinyaga and Muranga counties.

Kiambu County

Kiambu County borders Nairobi and Kajiado to the south, Machakos to the east, Murang'a to the North and Nyandarua to the Northwest. Kiambu is divided into upper Kiambu which is an extension of the Aberdare ranges and lower Kiambu. One of the theories suggest the name Kiambu originated from the word Kia mbuu meaning a place where screaming was common. Kiambu is one of the main coffees growing counties. It comprises both small holders and plantation farming. Some of the land in this county previously under coffee has now been sold for real estate lowering its production capacity.

County Profile

Altitude (meters above sea level)	Upper Kiambu 1800-2550 Lower Kiambu 1500-1800
Area (km^2)	2538.6
Population	2,417,735
Households	795,241
Population Density (persons per km^2)	952

KNBS 2019 (Kenya National Bureau of Statistics 2019)

Nyeri County

Nyeri County lies between the eastern base of the Aberdare range and western slopes of Mount Kenya. It borders Kirinyaga and Meru which are key coffee growing areas. The name Nyeri is said to have been derived from the Maasai word nyiro, given to the area by the Maasai people due to its red volcanic soil. Nyeri a predominantly coffee growing area is home to some of the best coffees in the world. Most of the coffee from this region is grown on small farms on the slopes of Mount Kenya.

County Profile

Altitude (meters above sea level)	1300-1900
Area (km^2)	3,325
Population	759,164
Households	248,050
Population Density (persons per km^2)	228

KNBS 2019 (Kenya National Bureau of Statistics 2019)

Kirinyaga County

Kirinyaga County lies at the foothills of Mount Kenya. Kirinyaga derives its name from Mount Kenya originally known as mount Kirinyaga. It borders Nyeri, Murang'a and Embu County. Kirinyaga undoubtedly produces some of the best coffees found in this region. It has an estimated 44,969 acres under coffee grown by about 67,578 farmers with 15 Co-operative Societies

County profile

Altitude (meters above sea level)	1300-1800
Area (km²)	1,478.3
Population	610,411
Households	204,188
Population Density (persons per km²)	413

KNBS 2019 (Kenya National Bureau of Statistics 2019)

Murang'a County

Murang'a County as known today was once known as Fort Hall. It's bordered by Nyeri, Kiambu, Nyandarua, Kirinyaga and Machakos. It is said that the name Murang'a originated from the word Kuranga. The Kikuyu people referred to going to the area for a visit as `Kuranga'. The name was later changed to Murang'a.

County profile

Altitude (meters above sea level)	1300-1800
Area (km²)	2,524.2
Population	1,056,640
Households	318,105
Population Density (persons per km²)	419

KNBS 2019 (Kenya National Bureau of Statistics 2019)

Eastern Kenya Region

Embu County

Embu County lies at the foothill of Mount Kenya. It borders Machakos to the south, Murang'a to the southwest and Kirinyaga to the west which are all coffee growing counties. It is named after the indigenous Aembu people.

County Profile

Altitude (meters above sea level)	1300-1800
Area (km²)	2,820.7
Population	608,599
Households	182,743
Population Density (persons per km²)	216

KNBS 2019 (Kenya National Bureau of Statistics 2019)

Meru County

Meru County borders Nyeri County to the southwest. The word Meru is believed to have originated from the word Mieru from the Maasai people, who referred to Tigania and Imenti forests in Meru County as Mieru forests meaning the quiet forests.

County Profile

Altitude (meters above sea level)	1300-1800
Area (km²)	7,006.3
Population	1,545,714
Households	426,360
Population Density (persons per km²)	221

KNBS 2019 (Kenya National Bureau of Statistics 2019)

Machakos County borders Nairobi, Kiambu, Embu, Kitui, Murang'a and Kirinyaga Counties. Machakos was the first administrative centre for the British colony in 1887 before moving it to Nairobi in 1899. It is named after Masaku who was an Akamba chief.

County profile

Altitude (meters above sea level)	1000-1600
Area (km^2)	6,042.7
Population	1,421,932
Households	402,466
Population Density (persons per km^2)	235

KNBS 2019 (Kenya National Bureau of Statistics 2019)

Makueni County is in the Ukambani region. It shares similar climatic conditions with Machakos County. Coffee is basically grown on the hilly areas of Iveti, Mbooni and Kangudo hills.

County profile

Altitude (meters above sea level)	1800-2002
Area (km^2)	8,169,8
Population	987,653
Households	244,669
Population Density (persons per km^2)	121

KNBS 2019 (Kenya National Bureau of Statistics 2019)

Western and Southwestern Kenya Region

Kisii County

Kisii County is found in southwestern Kenya. It has a high and reliable rainfall. The modern name Kisii is the corrupted version of Gusii, which is a shorter name for the Abagusii people who are the locals in this county. Kisii County grows Blue Mountain variety characterised by long oblong beans with mild acidity. Coffee in this region is mainly grown by small scale farmers.

County profile

Altitude (meters above sea level)	1450-1800
Area (km^2)	1,323
Population	1,266,860
Households	308,054
Population Density (persons per km^2)	958

KNBS 2019 (Kenya National Bureau of Statistics 2019)

Bungoma County.

Bungoma County is located in western Kenya. It borders the country of Uganda. Bungoma is said to have been a meeting place for the Bukusu elders who used drums "engoma' to summon people to meetings. Referred then to as a place of drums by the local Bukusu people it was named Bungoma. Coffee in Bungoma County is grown on the slopes of Mount Elgon.

The region is believed to have a high potential of being a major coffee growing County due the favourable coffee growing climatic conditions.

County profile

Altitude (meters above sea level)	1200-1950
Area (km²)	3,023.9
Population	1,670,570
Households	358,796
Population Density (persons per km²)	552

KNBS 2019 (Kenya National Bureau of Statistics 2019)

Rift Valley Region

Kericho County

Kericho County is named after a Masai chief Ole Kericho who was killed by the Gusii people in the 18th century. It is in the South Rift of the Great Rift Valley about 256 kilometres from Nairobi. Kericho is predominantly a tea growing county.

County profile

Altitude (meters above sea level)	1300-1800
Area (km²)	2,436.1
Population	901,777
Households	206,036
Population Density (persons per km²)	370

KNBS 2019 (Kenya National Bureau of Statistics 2019)

Trans Nzoia County

Trans Nzoia County is located on the western side of Mount Elgon. It is a multicultural county being home to more than four ethnic communities. It forms part of the Kenya food basket with commercial maize farming as the major agricultural activity besides coffee farming.

County Profile

Altitude (meters above sea level)	1700-1950
Area (km²)	2,495.2
Population	990,341
Households	223,808
Population Density (persons per km²)	397

KNBS 2019 (Kenya National Bureau of Statistics 2019)

Nakuru County

Nakuru County is in Kenya's Great Rift Valley which is multi-cultural. It is alleged the name Nakuru means a dusty place in Masai language. The name originated from frequent whirlwinds in the area accompanied by dusty clouds. Nakuru has a good climate that favours agriculture, with large scale farming, horticulture, dairy and coffee farming.

County Profile

Altitude (meters above sea level)	1750
Area (km²)	7,462.4
Population	2,162,202
Households	616,046
Population Density (persons per km²)	290

KNBS 2019 (Kenya National Bureau of Statistics 2019)

Baringo County

Baringo County is found in the former Rift Valley province. The name Baringo is derived from the word mparingo which means a lake in the local ethnic Njemps language. Coffee grown in this county is a on a small scale,

County profile

Altitude (Meters above sea level)	1067
Area (km²)	10,976.4
Population	666,763
Households	142,518
Population Density (persons per km²)	61

KNBS 2019 (Kenya National Bureau of Statistics 2019)

Coast Region

Taita Taveta County

Taita Taveta County is located northwest of Mombasa town, the main port in Kenya. Coffee is grown on a small scale in Taita Taveta County.

County Profile

Altitude (Meters above sea level)	500-2220
Area (km²)	17,152
Population	340,671
Households	96,429
Population Density (persons per km²)	20

KNBS 2019 (Kenya National Bureau of Statistics 2019)

CLEAN (GREEN) COFFEE PRODUCTION IN KILOGRAMS BY COUNTIES BETWEEN 2012/13 - 2018/19

COUNTY	2012/13	2013/14	2014/15	2015/16	2016/17	2017/18	2018/19
KIAMBU	11,783,596.75	10,000,299.67	11,345,200.00	12,769,604.22	10,910.799.79	7,527770.92	9,755,343.58
NYERI	5,228,601.50	9,159,251.50	5,656,458.00	7,524,940.62	4,088,390.92	5,030,604.12	5,317,897.24
KIRINYAGA	4,997,863.00	5,605,579.35	4,638,062.00	6,180,407.29	5,405,350.05	6,195,787.24	6,442,942.51
MURANGA	2,544,298.00	7,820,730.66	3,507,689.00	5,817,404.69	2,450,113.50	5,538,471.62	4,040,100.98
EMBU	2,291,172.75	2,404.233.78	2,003,363.00	2,071,095.92	2,057,774.13	2,374.282.42	2,246,568.07
BUNGOMA	2,035,668.00	1,881,838.15	1,247,335.00	1,286,671.05	1,763,331.09	2,317,019.59	2,179,570.23
MERU	1,973,442.40	2,744,449.67	3,974,854.00	1,727,242.59	1,646,899.51	1,727.051.60	1,593,002.06
KERICHO	1,813,456.00	1,294,226.76	1,829,221.00	1,444,850.71	2,730,804.52	2,199.905.61	3,857,672.11
MACHAKOS	1,650,424.50	2,899,787.77	2,329,268.00	1,950,724.48	1,600,370.61	1,387,875.89	2,553,554.90
KISII	1,264,076.90	1,188,555.89	1,091,783.00	1,412,009.97	1,099.743.18	2,433,899.92	1,420,346.11
MAKUENI	1,005,794.00	197,883.55	174,286.00	67,365.85	102,476.88	88,977.29	108,894.12
THARAKA-NITHI	963,600.00	1,493.840.79	1,385,575.00	861,541.29	945,170.01	834,894.07	1,386,233.85
NYAMIRA	824,071.00	1,001,381.93	886,808.00	1,016,590.69	916,394.60	1,583,417.97	774,082.82
TRANS NZOIA	482,696.00	426,572.53	641,898.00	104,707.41	998,433.33	658,307.61	749,578.82
NAKURU	444,440.00	688.575.49	626.181.00	797,759.96	779,697.24	353,054.59	943,544.25

COUNTY	2012/13	2013/14	2014/15	2015/16	2016/17	2017/18	2018/19
NANDI	225,978.00	256,283.14	418,676.00	383,010.09	584.620.82	688,740.66	848,893.46
UASIN GISHU	40,262.00	9,768.26	17,950.00	47,199.00	28,010.64	18,585.49	55,342.49
MIGORI	26,444.00	94,035.33	40,440.00	75,107.29	70,333.42	105,439.59	153,969.61
HOMA BAY	26,176.00	49,082.93	31,453.00	32,026.90	60,341.00	37,815.00	63,299.00
BOMET	13,906.00	9,956.37	8,413.00	13,509.33	27,974.76	4,258.07	27,579.36
BARINGO	13,523.00	88,406.88	89,311.00	77,735.39	120,197.43	65,216.21	135,510.40
WEST POKOT	11,055.00	29,116.87	30,928.00	17,295.00	39,286.27	25,472.00	30,469.00
ELGEYO MARAKWET	10,758.00	16,180.72	21,759.00	20,837.00	54,792.00	45.746.00	58,893.00
LAIKIPIA	8,283.00	4,481.19	8928.00	7324.00	14,097.44	6,916.59	4,809.50
NAROK	1,463.00	1,191.29				47.48	2,160.76
KISUMU		50,086.41	3,318.00	2,614.61	12,632.00	26,320.00	69,830.10
KAKAMEGA	7,598.95	28,085.44	26,813.00	15,042.81	26,840.87	25,587.26	29,018.00
KAJIADO	-	4,234.96	-	-	-	-	1,645.00
BUSIA	-	3,673.99	-	4,590.00	-	-	11,346.21 (Robusta)
TAITA		3,292.79	1,770.00	1,800.45	878.36		-
VIHIGA	64.05	23.71	-	-	349.00	1,373.00	560.00
NAIROBI	-	-	-	-	83,990.59	99,172.00	132,718.00
TOTAL	**39,685,711.80**	**49,475,320.77**	**42,037,740.00**	**46,121,376.61**	**38,620,093.98**	**41,375,009.79**	**44,984,029.32**

Source: Agriculture and Food Authority (Coffee Directorate) coffee yearbook season 2018/19

The table above shows the coffee production by each county from 2012/13 to 2018/19

CLEAN (GREEN) COFFEE PRODUCTION IN KILOGRAMS BY ESTATES AND COOPERATIVE SOCIETIES BY COUNTIES IN 2018/19 (ARABICA AND ROBUSTA)

NO	COUNTY	ESTATES (Large farms)	CO-OPERTIVE SOCIETIES	TOTAL Kilograms
1	KIAMBU	8,639.229.38	1,116114.20	9,755,343.58
2	KIRINYAGA	426,194.36	6,016,748.15	6,442,942.51
3	NYERI	831,726.46	4,486,170.78	5,317,897.24
4	MURANGA	716,385.65	3,323,715.33	4,040,100.98
5	KERICHO	346,682.46	3,510,989.65	3,857,672.11
6	MACHAKOS	361,029.37	2,192,525.53	2,553,554.90
7	EMBU	176,995.05	2,069,573.02	2,246,568.07
8	BUNGOMA	49,760.29	2,129,809.94	2,179,570.23
9	MERU	174,343.73	1,418,658.33	1,593,002.06
10	KISII	492,165.19	928,180.92	1,420,346.11
11	THARAKA NITHI	100,895.41	1,285,338.44	1,386,233.85
12	NAKURU	741,854.25	201,690.00	943,544.25
13	NANDI	26,854.33	822,03913	848,893.46

NO	COUNTY	ESTATES (Large farms)	CO-OPERTIVE SOCIETIES	TOTAL Kilograms
14	NYAMIRA	214,354.59	559,728.23	774,082.82
15	TRANS NZOIA	519,979.51	229,599.31	749,578.82
16	MIGORI	3,540.95	150,428.66	153.969.61
17	BARINGO	26,478.40	109.032.00	135,510.40
18	NAIROBI	132,718.00		132,718.00
19	MAKUENI	16,637.00	92,257.12	108,894.12
20	KISUMU	61,582.10	8,248.00	69,830.10
21	HOMA BAY		63,299.00	63,299.00
22	ELGEYO MARAKWET	5,992.00	52,901.00	58,893.00
23	UASIN GISHU	31,151.00	24,191.49	55,342.49
24	WEST POKOT		30,469.00	30,469.00
25	KAKAMEGA	5,119.00	23,899.00	29,018.00
26	BOMET	78.34	27,501.02	27,579.36
27	LAIKIPIA		4,809.50	4,809.50
28	NAROK		2,160.76	2,160.76
29	KAJIADO	1,645.00		1,645.00
30	VIHIGA		560.00	560.00
31	BUSIA-(ROBUSTA)		3000.00	
	GRAND TOTAL	**14,103,391.82**	**30,880,637.50**	**44,984,029.32**
	PRODUCTION %	**31.36 %**	**68.64 %**	**100 %**

Source: Agriculture and Food Authority (Coffee Directorate) coffee yearbook season 2018/19

This table compares production between the Cooperative Societies and Estate (Plantations) coffee sectors in Kenya in the 2018/19 coffee year. Co-operative Societies still supply the bulk of Kenyan coffee.

CHAPTER 5

FACTORS THAT FAVOUR COFFEE FARMING IN KENYA.

Red Volcanic Soil:

Arabica coffee does not grow in water-logged soil. It requires well drained soil. The holes dug for the coffee tree planting must be at least 3 metres deep to enable the coffee roots to extract water in times of drought. Kenya has volcanic soil even deeper than 3 metres in some areas making it ideal for coffee growing.

Volcanic soil is especially found around Mount Kenya where the best of Kenya coffee is grown

Photo courtesy of Thunguri factory

3ft x3ft x 3ft hole for planting a coffee seedling. It is filled with leaves, sub soil and water

High rainfall:

Kenya coffee growing areas receive well above 1000 millimetres annually which is generally well distributed around the year. This amount of rainfall is required for flowering, growing and maturing quality coffee cherries that produce regular solid beans. It is also true to say that Kenya also occasionally faces drought, a challenge across the globe. Some estates use irrigation and are therefore able to mitigate drought conditions

Coffee Irrigation

Optimum temperature

The Temperature of the Kenya coffee belt averages between 15°C to about 24°C. Coffee growing areas especially around Mount Kenya experience ideal temperature conditions. Lower temperatures found in high elevated areas, are believed to slow fruit maturation enhancing the inherent quality of the coffee beans through prolonged flavour development, a reason as to why some of the best coffee of the world comes from this region. Some lower altitude growing regions like Machakos experience slightly higher temperatures

Altitude*:*

Kenyan *c*offee is grown in areas with an altitude of between 1400m to 2200m above sea level. The altitude influences both the rainfall and cooler temperature conditions. High grown and shade grown coffees are believed to mature slowly developing full flavour potential. These coffee beans acquire good inherent quality producing some of best coffees in world with a unique cup profile.

Coffee Tree Husbandry

Kenya coffee is grown over five to six-year cycles. At the end of the cycle the trees are cut to the stumps to encourage growth of new trunks.

photo courtesy of Thunguri factory

An over 60-year-old coffee tree due for replacement

Coffee trees stand at less than three meters high to allow picking and pruning. Pruning takes place three to four times in a year to maintain an optimum leaf to bean ratio.

One of the benefits of pruning is to ensure the tree does not use all its resources growing leaves at the expense of cherry production. Pruning also allows light to permeate through.

Standardised expertise:

The Coffee Directorate certifies all coffee experts (tasters) in collaboration with Coffee Traders' Association through a structured programme. Successful candidates are issued with a license.

These trained coffee experts work in the milling, export, and the marketing sectors. Part of their role is to advise the farmers on quality related issues. The Kenya Coffee Research Institute through the Coffee College also conducts training using tailor made programmes for farm managers, farmers and factory managers on coffee husbandry and processing.

It is a mandatory requirement for all companies in the coffee industry to have a licensed quality expert.

Kenyan farmer's tenacity and consistency in taking care of their coffee through different price cycles eventually pays off. Many farmers who do not abandon their coffee during price fluctuations have better earnings in the long run.

During the quota system in the 80s and 90s it was a prestigious role to be a coffee farmer as they were economically above the general populace.

Photo courtesy of Thunguri factory

Chairman Thunguri factory Mr Kanyi Pruning his coffee tree

Premium Status of Coffee

Kenyan coffee generally enjoys premium prices compared to coffee from other origins. This encourages farmers to strive to maintain a premium quality coffee

CHAPTER 6

COFFEE SPECIES GROWN IN KENYA

Coffee species

There are two main coffee species grown worldwide, namely Arabica and Robusta. Arabica is mostly dominant in Central and South America and much of East Africa.

Bourbon and Typica are the main culturally and genetically important groups of Arabica coffee in the world. The World Coffee Research Organisation estimates that more than 80 % of Arabica coffee production worldwide derives from Typica and Bourbon related varieties. Typica is a Latin word meaning ordinary and is planted all over the world.

Bourbon was named after an island in the Indian Ocean currently called Reunion. Its unusual story began in 1715, when the coffee plants originating in Ethiopia were brought from Yemen to the Bourbon Island then under France control to be cultivated there. The plants were said to have thrived very well and slowly through mutation acquired their own identity. The French mission then imported this Bourbon species to East Africa.

Kenya produces about 99% Arabica and about 1% of Robusta coffee. Land under coffee production is estimated at 115,570 hectares. Of this total 88,278 hectares fall under cooperatives while 26,222 hectares falls under large plantations (Estates). (ICO-ICC 124-7)

Robusta coffee grows in Western Kenya and some parts of the coast region. Central and Eastern Kenya regions account for about 80 % of the total Arabica coffee produced in Kenya, while 20 % is grown in the Western, Coast and Rift Valley regions. Most of the best coffee that Kenya is known for is grown around the Mount Kenya region.

Kenya Arabica Coffee Main Coffee Varieties.

All the traditional commercial Arabica varieties namely SL28, SL34, and K7 have their lineage from the French Mission coffee, which is a Bourbon derivative.

History and characteristics of Kenya commercial Arabica coffee varieties.

SL 28

SL 28 was selected in 1935 at the former Scott Agricultural Laboratories from a single tree in a population known as Tanzania Drought Resistant. Scott Laboratories named after Dr Henry Scott of the Church of Scotland Mission was established in a building formerly used as a sanatorium hospital in Kikuyu town.

This variety is bourbon related and one of the best varieties in Africa. It's more suited to medium to high attitudes and bears a bold bean. At its full potential SL28 has a cup with black currant and raspberry notes. The prefix SL at the beginning of the variety name stands for Scott Laboratories. According to the "World Research Organisation Varieties Catalogue", this variety is also found in Malawi, Uganda, and Zimbabwe.

Historical documentation has it that A.D.Trench, a senior coffee research officer at Scott Laboratories while conducting a tour in Tanzania Moduli district noticed a variety which appeared tolerant to drought, diseases and pests. He brought the variety back to Scott Laboratories where its resistance to drought was confirmed and it was subsequently widely distributed.

Its non-susceptibility to diseases, allows the tree to be left unattended for years and then returned to successful production. There are SL 28 trees that are 60 to 80 years old that are still productive.

SL 34

SL 34 is a Bourbon variety and produces at its best quality an excellent cup next to SL28. This variety grows at high altitude areas with high rainfall and has a high yield. SL 34 was selected at the Scott Laboratories in 1930s from a single tree from the Loresho Estate in Kabete. It was among individual trees selected between 1935-1939 and prefixed SL. Recent genetic tests suggest SL 34 is related to the Typica genetic group

K7

Bourbon related K7 is found in Kenya and Tanzania. This variety is more suited to low altitude areas where leaf rust is prevalent. K7 was selected in 1936 from French Mission trees from the Legetet Estate Muhoroni in Kenya. Produced in low quantity, its cup quality is not as good as SL 28 and SL 34 cups. Although K7 was immune to coffee leaf rust for several years this is no longer the case.

Ruiru11

This is a composite variety that was released in 1985 and is best suited in areas with a high coffee berry disease infestation and leaf rust. The name Ruiru is derived from the location of the Kenya Coffee Research Institute Station where it was developed. It produces bold beans with a balanced cup with low to medium acidity and medium body compared to other Kenyan varieties with higher acidity and complex flavour.

The eventual development and introduction of Ruiru 11 was in response to a coffee berry disease epidemic in 1968 that wiped 50 % of Kenya production. According to the "World Coffee Research Varieties Catalogue"in the composite of Ruiru 11 variety is Catimor which has some Robusta genes. Catimor is a hybrid of Caturra and Timor. Timor is a hybrid of Robusta and Arabica.

Batian

The Batian variety is named after the third peak of Mount Kenya which is named after a prominent Masai leader. Released in May 2010 by the Kenya Coffee Research Institute, Batian is a composite variety that produces bold beans and is a high yielding variety. Batian at its full potential has a particularly good cup with good acidity, medium body, and a sweet and complex flavour. According to the Kenya Coffee Research Institute it is resistant to coffee leaf rust and coffee berry diseases providing a significant reduction in production cost.

Coffee Seedling nursery

Other coffee Varieties

Blue Mountain

This variety is as old as the French mission. It's known for its appearance, which is characterized by a softish oblong bean that tends to slightly curve inwards. It is prone to damage by insects and normally has a cup profile with a mild coffee flavour, and low acidity. This variety is grown in the Kisii region and some parts of Western Kenya. The Blue Mountain variety was imported from Jamaica and planted around 1913 but was not successful in other parts of Kenya.

Other varieties include Kent grown in Meru and Solai in Nakuru. Solai is now almost extinct. Most of the old plantations cultivate a combination of varieties, including SL 34, SL 28, K 7 and Ruiru 11.

CHAPTER 7

KENYA COFFEE INDUSTRY STAKE HOLDERS

Categories and Numbers of registered stakeholders with AFA (Coffee Directorate).

No	Category	Number of entities
1	Smallholders Farmers	700,000
2	Estates	3000
3	Co-operative Societies	500
4	Co-operative Unions	14
5	Co-operatives owned pulping (washing) stations	1001
6	Estates owned pulping (Washing) stations	3000
7	Coffee millers	17
8	Commercial agents	10
9	Growers Marketing agents	25
10	Coffee exchange	1
11	Coffee dealers	77

Source: ICO (Kenya Coffee Profile)

The bulk of coffee grown is by smallholder farms, grouped into co-operative and estates. There are an estimated 800,000 Smallholder farmers grouped into 500 co-operative societies. There are 3000 coffee estates categorised as small, medium, and large.

Distribution of Coffee Holdings

Sector	Size of acreage	Number of farmers	Share of total farmers
Small holder affiliated	Varies < 5 acres	800,000	99.63 %
Small estates	5-20 acres	2400	0.30 %
Medium estates	20-50 acres	500	0.06 %
Large estates	Over 50 acres	100	0.01 %
Total number of coffee farmers		803,000	100 %

Source: ICO (Kenya Coffee Profile)

Most coffee grown in Kenya is from small holding.

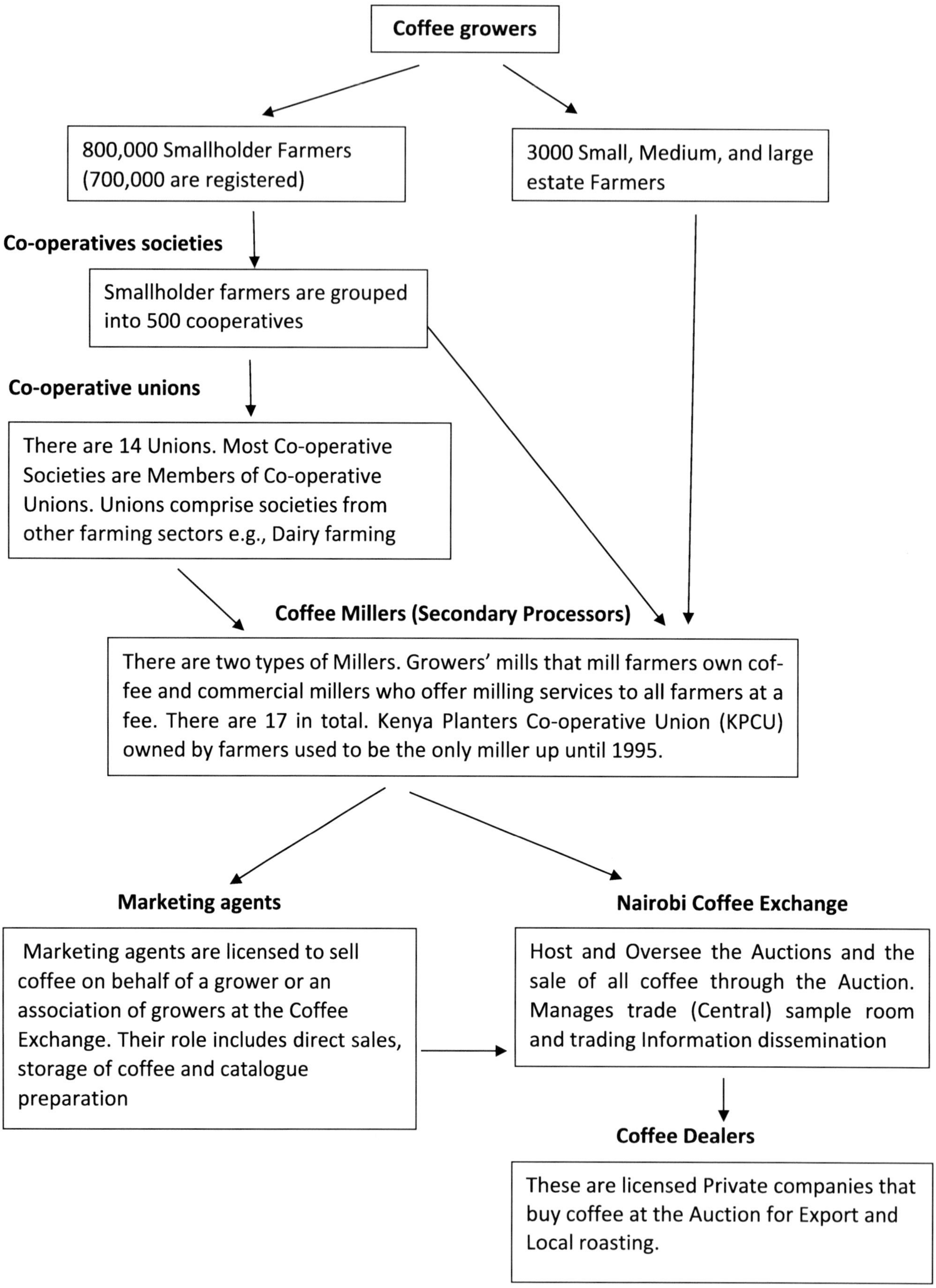
Coffee growers
800,000 Smallholder Farmers (700,000 are registered)
3000 Small, Medium, and large estate Farmers
Co-operatives societies
Smallholder farmers are grouped into 500 cooperatives
Co-operative unions
There are 14 Unions. Most Co-operative Societies are Members of Co-operative Unions. Unions comprise societies from other farming sectors e.g., Dairy farming
Coffee Millers (Secondary Processors)
There are two types of Millers. Growers' mills that mill farmers own coffee and commercial millers who offer milling services to all farmers at a fee. There are 17 in total. Kenya Planters Co-operative Union (KPCU) owned by farmers used to be the only miller up until 1995.
Marketing agents
Marketing agents are licensed to sell coffee on behalf of a grower or an association of growers at the Coffee Exchange. Their role includes direct sales, storage of coffee and catalogue preparation
Nairobi Coffee Exchange
Host and Oversee the Auctions and the sale of all coffee through the Auction. Manages trade (Central) sample room and trading Information dissemination
Coffee Dealers
These are licensed Private companies that buy coffee at the Auction for Export and Local roasting.

Smallholder coffee growers

Smallholder farmers own small farms less than five acres and generally produce high-quality coffees. These small farms are owned by individual family units who provide the labour required.

Smallholder farmers are organized into Coffee Co-operative Societies who sell the coffee on the farmers behalf.

The Co-operative Societies receive cherries for primary processing and deliver the parchment to the Millers on behalf of their members for milling and finally direct sale or auction.

Ripe cherries ready for picking

Co-operative societies enable individual farmers to corporately own a factory (pulping or washing station) cutting down on production costs. The largest quantity of high-quality coffee is produced by cooperatives societies.

Manual labour at the factories is provided by society members (farmers) and their families due to limited resources

In the early days all co-operative societies were affiliated into district unions. Most district unions affiliated with coffee societies also managed other cooperative societies like dairy.

Estate coffee growers

There are 3000 coffee estates ranging from small, medium, and large. Small Estate farmers own farms between five to twenty acres. Medium estates own farms between twenty to fifty acres while large estates own more than fifty acres.

These Estates can acquire licenses to operate their own pulping (washing stations) for primary processing of cherries before delivering the parchment for milling to the Dry mills. Some large estates also own their own Dry Mills. Those with commercial licenses can mill other farmer's coffee at a cost.

Kenya coffee hulled on the farm is known as "estate cured" coffee and is sold either as UG (ungraded coffee) or graded to standard grades. Coffee estates (plantations) use both manual and mechanised processes. Most Coffee Estates produce high volumes of average quality with few producing high-quality similar to co-operative coffees.

An Estate, a large size coffee farm.

Coffee Millers

Coffee millers are secondary coffee processors who mill parchment coffee, including Mbuni (dry processed coffee), and grade estate cured coffee.

Coffee dealers (Exporters)

Coffee dealers buy clean coffee (green beans without parchment) from the auction for export or local secondary market.

Coffee Grower Marketers

A Grower Marketing License allows farmers to market their own coffee outside the auction system through direct sales

Coffee Marketing Agents

Marketing agents sell (green beans) coffee at the auction on behalf of the growers at a fee. Their main service includes auctioning of coffee, handling, contract documents preparation and warehousing.

Coffee is delivered to the Marketing Agents after milling. They evaluate the quality of every lot of coffee and prepare the sale catalogues on a weekly basis for the auction. They draw samples from all the lots offered in each sale and avail them to buyers in the trade sample room.

Coffee warehousing and storage

Operators who own warehouses must have a Warehousing License to store coffee on behalf of other coffee stakeholders at a fee. Coffee is only allowed to be stored in these warehouses.

Nairobi Coffee Exchange

Nairobi Coffee Exchange is an organisation which hosts and oversees weekly coffee auctions. The Central Auction is currently run by a management committee comprising members drawn from the Coffee Stakeholders who may include farmers, millers, and traders. The Exchange being a non-profit organization is funded by membership fees and levies on the sale of the coffee from traders who participate in the coffee auction.

CHAPTER 8

HISTORY AND EVOLUTION OF THE KENYA COFFEE MARKETING SYSTEM

The Kenya coffee marketing system has evolved over the years in a bid to combat emerging challenges facing farmers and the industry at large.

Under the International Coffee Agreement, Kenya internally marketed its coffee through the "Pool System" which was aligned to the International Quota System.

The Quota System introduced in 1962 collapsed in 1989. Quotas were allocated based on quality and coffee groups which included Colombian Mild, (Kenya Colombia and Tanzania) and Other Mild's Arabicas with the sole purpose of avoiding market gluts ensuring stability in coffee prices.

Kenya coffee was categorised under the "Colombian Mild's", a top echelon coffee in the market. This group was assigned by the International Coffee Organisation and used to allocate quotas based on quality among similar world producers.

Through the Pool System, farmers were paid in instalments even before their coffee was sold or partially sold.

POOL SYSTEM

The pool system of marketing coffee was the initial internal established system used by the then Coffee Board of Kenya responsible for the marketing of all the coffee in the country. A pool was created for all the coffee delivered within a given coffee production year. This system guaranteed farmers subsequent payments based on their quantity and quality after milling and sale within a pool year.

The payments were made in intervals between the time the coffee was delivered to the mills through to the end of the coffee production year. This was after the deduction of marketing expenses and statutory deductions by the then Coffee Board of Kenya. It was the only option at the time for marketing coffee.

The Pool System assured farmers of subsequent payments throughout the pool year. Farmers without other sources of credit facilities and who did not want to incur high interest on loans could rely on the system for cash flow. Moreover, all the farmers in the pool shared the risk of price fluctuations within that production year. This meant irrespective of the quota allocated to Kenya in a given production year all the farmers got paid.

Payments associated with the Pool System were made in small amounts and at times the payment intervals could be stretched out into long intervals. The farmer could not determine the timing of the payments. Quality was based on a classification system with classes ranging from 1 to 10.

This classification system was based on the visual appearance of both the green and roasted beans and the cup. Farmers whose coffee had unique flavour notes and complexity with higher cup scores were paid at a similar rate determined for that class category. Farmers with coffee that earned a premium shared these benefits with other farmers in the pool.

Price differentials between different lots within the same class or quality range were common but payment to the farmers was based on the price of the class category.

Another disadvantage of the pool system was the bulking of big lots of coffee for milling and marketing compromised coffee uniqueness and identity in terms of any special flavour profile notes and quality value associated with an individual lot.

Payments in a Pool Year.

Parchment Advance (First payment)

This payment was made on application by a farmer or a co-operative society through their agent to the board. It was based and guaranteed by coffee already picked at a rate determined by the board and included interest payments on money borrowed. The rate was paid per bag of parchment.

Part Payment Advance (Second Payment)

This payment was made after parchment coffee was delivered to the Mill operated by the Kenya Planters Co-operative Union (KPCU) which at the time had the monopoly for milling all the coffee in the country. The rate for payment was determined by the Coffee Board and was based on parchment category whether parchment grades one, two, three or lights. Part payment advance (2nd payment) did not include an interest charge and was deducted from the initial Payment (3rd payment).

Initial Payment (Third payment)

The board was required under the Coffee Act to make an initial part payment (3rd Payment) to the farmers within fourteen days from the date when their coffee was classified (evaluated and quality determined). The payments were made weekly at a rate determined by the board at the beginning of the calendar year based on the class of coffee ranging from Class 1 to10. Coffees of the same class were paid the same rate, influenced by prevailing world prices at the time.

Interim Payments

Interim part payments were an upward revision of the initial payment and were done during the year as the cash flow improved together with higher confidence of world price levels being achieved. The rate for interim payments was governed by the prices prevailing in the market and the cash flow of the board. The frequency of interim payments was determined by price trends. Upward price trends meant more interim payments, while a downward price trend meant fewer interim payments. These payments were based on cumulative kilos of classified coffee for a given farmer.

Final Payment

The final payment was the balance owed to the farmer in a year after providing for the Coffee Board's marketing expenses. The payment was prepared after stocks had been valued and accounts prepared. Final payment for each class was the difference between the annual average payment and the amount already paid during the coffee production year.

Pool Payment Model Flow Chart

Coffee payment to Individual farmers was not dependent on whether their coffee had been sold at the **Auction.**

↓

Parchment Advance (1st Payment)
Secured by parchment or already picked cherry

↓

Part Payment Advance (2nd Payment) was made based on the category of parchment delivered to the mill with varying rates for parchment grade one, two, three and lights.

↓

Initial Payment (3rd Payment) was paid within 14 days after classification of a given coffee outturn (delivery) by the coffee board based on the class of the coffee. Classes ranging from 1-10

↓

Interim payment was an upward revision of the Initial payment determined by cash flow and Coffee world price movement

↓

Final Payment was made after completion of milling and all the coffee commercially valued.

Pool Marketing System Value Chain

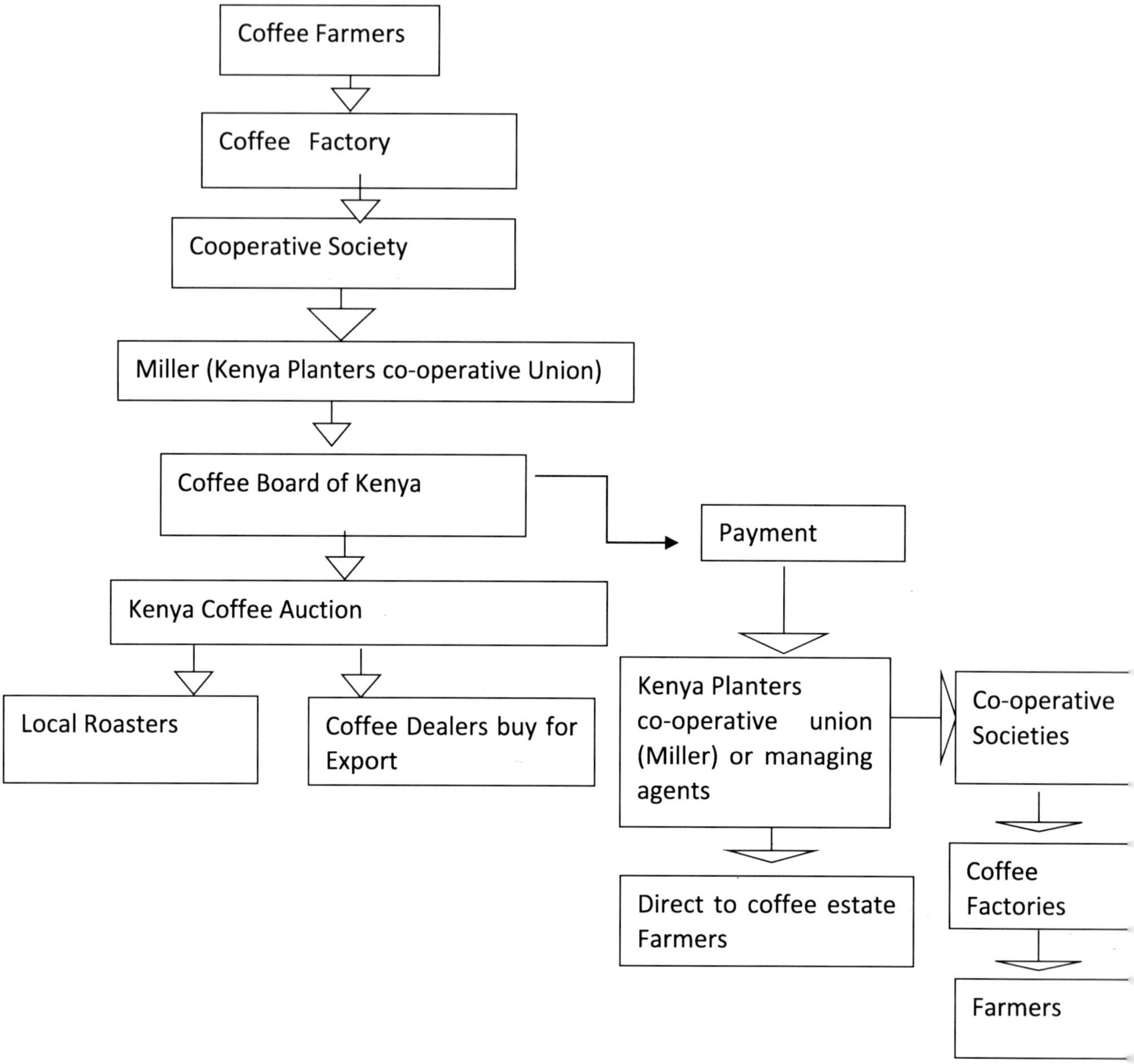

First phase of Coffee Marketing Reforms

Trading Currency

Reforms in the coffee sector started 27th October 1992 with the introduction of the auctioning of coffee in US dollars from Kenya shillings, the currency that coffee is traded in today.

Following this change, the Board initially was only allowed to retain 50 % of the coffee proceeds in US dollars with the remaining 50 % sold to the Central Bank of Kenya at the official exchange rate. This meant that the Coffee Board of Kenya had to carry the payments in both Kenya shilling and US dollars on a 50 % basis.

Further review of this policy allowed the Board to retain 100 % US dollars from the sale proceeds. Payment to the farmers from the 15th of February1994 were made in US dollars. Trading of coffee in US dollars meant that farmers could be cushioned from fluctuation of the Kenya Shilling.

Coffee Marketing outside the Pool

This was a marketing option, where farmers were allowed to auction their coffee outside the normal pool system. Coffee outside the Pool was sold, and payment made in full based on value of the coffee minus statutory and marketing expenses. The payment was based on weight for an Outturn Number delivered with no further payment for the same consignment afterwards.

Coffee sold outside the pool was paid in full after the sale unlike coffee sold in the pool. This system rewarded farmers whose coffee fetched special premiums in the market, therefore encouraging production of high-quality coffee. This system did not allow for part payments unless a farmer sold his coffee in both the Pool and outside Pool channels. This meant if the coffee took a long time before it was sold the farmer could not access any advance payment within this time.

Marketing coffee outside of the Pool system exposed the farmer to the risk of 100% price fluctuation

Outside pool Payment model

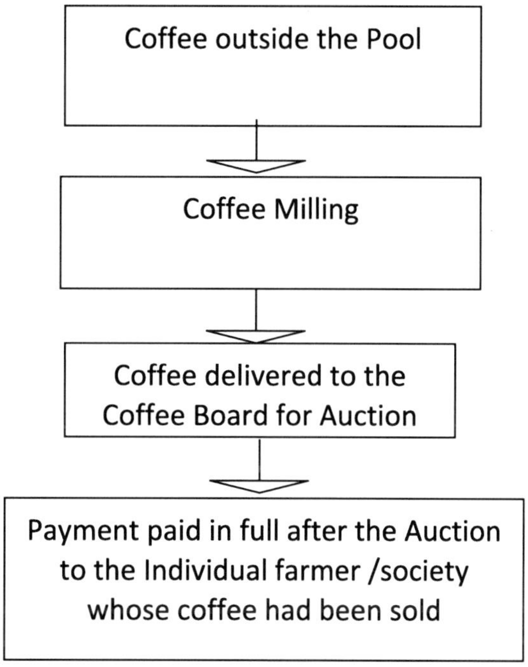

Licensing More Millers

In 1995 more Commercial Millers were licensed ending the monopoly of the Kenya Planter's Cooperative Union Mill. The Mills offered their services at a fee.

At this time the Kenya Planters Co-operative Union milled coffee mainly from Co-operative Societies and had the largest market share of the coffee milling sector. The distribution of the milling market share continued to change as farmers moved from one miller to another creating stiff competition among millers. This competition increased due to limited and declining coffee production compared to available milling capacity.

Multi auctioneering

Since the creation of the Coffee Board of Kenya, the Kenya Coffee Auction was the sole broker selling coffee on behalf of the Board. On 1st October 1997, the Board introduced a multi auctioneering system ending the monopoly of Kenya Coffee Auction which was jointly owned by the Coffee Board of Kenya and the Valentine family, who ran the auction. This change was designed to give farmers alternative Auctioneers. With the introduction of this change each licensed commercial miller formed their own subsidiary to auction coffee on behalf of their farmers. Multi auctioneering was meant to encourage competition and deliver competitive fees and better services to farmers.

Second Phase Coffee Marketing Reforms

Multiple Marketing agents

Prior to reforms in the coffee sector, the Coffee Board of Kenya (now the Coffee Directorate) doubled up as a regulatory body and the sole marketing agent for all the coffee in Kenya. With reforms in the industry, its role was reduced to a supervisory and regulatory body for the coffee industry.

In 2001 the government appointed three interim marketing agents all who happened to be commercial millers, opening the coffee marketing agency to other players. A Marketing Agent's role included promoting Kenyan coffee in the local and international market. They ensured coffee delivered to them was sold at the best price possible while maintaining the flow of coffee supply to buyers through the auction. Previously all the coffee was handed over to Coffee Board of Kenya who would then offer the coffee for sale at the Nairobi Coffee Exchange.

All the Marketing Agents at the time were subsidiaries of commercial millers. Reforms of the coffee marketing structure created two types of marketing agents

A **Grower Marketing Agent** who was licenced to sell their own coffees through the auction process or directly to overseas buyers. This category comprised of coffee farmer's unions, societies and estate owners or associations. Introduction of direct sales was to provide farmers with an alternative marketing channel

Commercial Marketing Agents were licensed to market coffee either through the auction process or directly to overseas buyers on behalf of the farmers. They were required to prepare coffee sales catalogues and offer coffee at the Nairobi Coffee Exchange through their auctioneers.

Once coffee was sold, they would remit the proceeds to the farmer, either directly or through their society, after deduction of the marketing charges.

Direct Sales (Second Window)

On the 21st July 2006, the government as part of its reform agenda introduced another marketing channel dubbed the 'Second Window'. This channel allowed Kenyan farmers to bypass the auction and sell their coffee directly to overseas buyers.

Direct sale of green bean only (clean coffee) involves a contractual agreement being established between an overseas buyer and a Grower, Co-operatives Society, Union, Grower Miller, Estate, or an association of coffee growers.

There is a proposal for further changes to Direct Sale Contracts to allow local sales previously prohibited. Ownership of the coffee is not passed on to the buyer until it has been paid for. The price for Direct Sale coffee must be above the auction price for a similar quality.

One of the challenges of a Direct Sale Contract over a long period, is guaranteeing the buyer's consistency in quality and quantity over the contractual period. Consistency in quality and quantity is subject to several external factors including rainfall distribution, frost and drought that are not within the producer's control. Coffee sold through Direct Sale is on a small scale compared to coffee sold at the auction, although it is growing. Prices are subject to demand and supply both at the auction and Direct Sales.

The Auction Price acts as the benchmark for Direct Sale prices highlighting the importance of the Central Auction System.

Most of the coffee sold through Direct Sale, targets the speciality niche market for high quality coffee which commands a premium price. Although growing the speciality market is a small percentage of the entire coffee market. This limits the impact of Direct Sale prices on the entire industry. Higher prices associated with the speciality market encourage production of high-quality coffee.

Direct sales vs Auctioned Coffee between 2007/2008 to 2017/2018

Coffee Year	No of Kilograms sold through the Auction	No of Kilograms sold through Direct sale	Total Kilograms	% Of direct sales
2007/2008	39,461,673	2,045,280	41,506,953	4.92 %
2008/2009	51,881,026	1,195,560	53,076,586	2.25 %
2009/2010	36,197,159	3,784,154	39,981,313	9.46 %
2010/2011	33,633,235	2,231,981	35,865,216	6.22 %
2011/2012	43,366,142	5,701,514	49,067,656	11.61 %
2012/2013	38,141,308	3,723,449	41,864,757	8.89 %
2013/2014	40,927,603	8,282,060	49,209,663	16.83 %
2014/2015	34,754,166	7,183,000	41,937,166	17.12 %
2015/2016	38,375,591	7,543,758	45,919,349	16.42 %
2016/2017	34,090,505	6,755,743	40,846,248	16.53 %
2017/2018	35,675,449	7,962,253	43,637,702	18.24 %

Source: ICO (Kenya Coffee Profile)

This Table compares the amount of coffee sold through the Auction and direct sales between 2007 and 2018

Second phase Coffee Reforms Marketing Value Chain

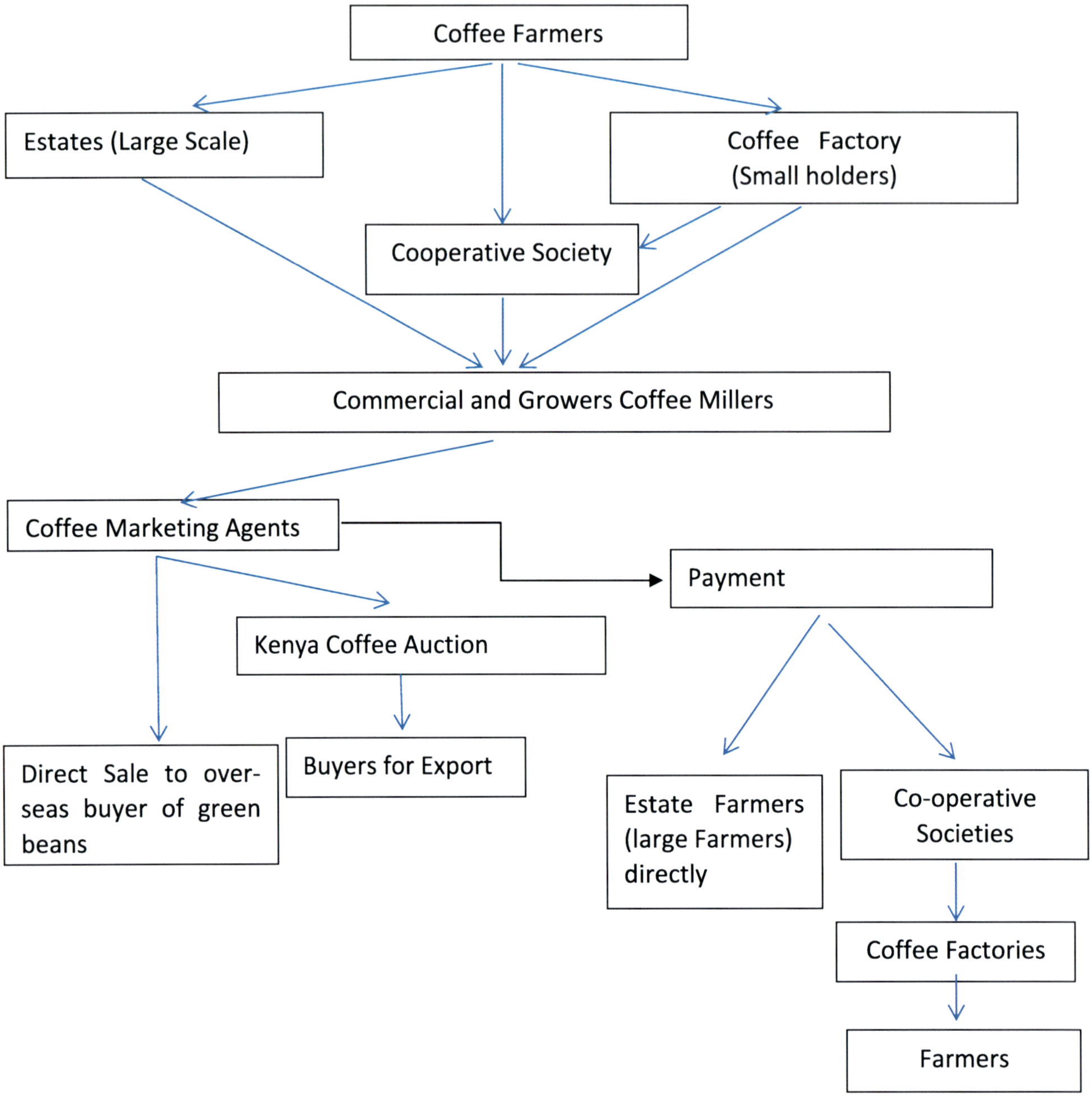

Proposed Third Phase Kenya Coffee Reforms.

The government is proposing further changes to the coffee industry through the Coffee Bill of year 2020. Among the proposed changes is a single license category for an individual player in the industry, to eliminate conflict of interest among the players. Unlike previously where one company or the same directorship held multiple licenses for example in milling, export, and local roasting, under proposed new regulation a player will only hold one license category.

Key in these reforms is reserving a pulping license only for farmers as has been previously the case. It proposes analysis of coffee parchment before milling to determine the anticipated milling loss.

Another proposal is the introduction of cupping centres to provide farmers with an independent report apart from the miller's report. A similar system existed during the pool system where the same coffee was classified by both the Kenya Planter's Cooperative Union and the Coffee Board of Kenya with a provision for appeal by farmers against the Coffee Board classification results.

Grower's Miller license holders like licensed Brokers will be allowed to trade at the auction. They can also decide to use the services of a Broker to trade at the auction. Growers, Co-operative Societies, unions, grower's millers, estate, or an association of coffee growers can directly sell their coffee in green (clean) form in the local market to a roaster or an overseas buyer through a contractual agreement.

The Nairobi Coffee Exchange will continue to manage the central sample room, information registry and the direct settlement scheme. It will also maintain records relating to coffee sales and sweepings and avail catalogues to relevant parties.

In the new proposal, Cooperative Unions (which constitute various Co-operative Societies) or any legal licensed entity representing the farmer, can participate in direct sales of their own green coffee.

While farmers have a right to determine their destiny, the government should midwife any changes to avoid confusion in the industry and conflict among stake holders. Changes that do not add any value to farmer's earnings or enhance efficiency should be discarded.

The role of the County Governments in the industry should be well defined to avoid duplication of roles under the National Government, to reduce business operational costs and unnecessary bureaucracy.

CHAPTER 9

COFFEE FLOWERING AND HARVESTING

Photo courtesy of Thunguri factory

Flowering of the Fly crop (early crop) occurs between September and October with harvesting taking place between May and July. The main crop (late crop) flowering occurs between February and March while harvesting occurs between September to December. It takes about six months from flowering to harvesting. The Fly crop gets to the auction around August to December, while the main crop gets to auction from around December to May.

The auction spread, and quantities offered at auction depend on the current crop quantities. Under normal circumstance the auction is scheduled weekly as earlier mentioned. Flowering and harvesting can be slightly delayed due to changes in weather patterns.

The Fly crop (otherwise known as "early crop) as the story goes was a term coined from the speed at which the pickers harvested the crop. Cherries are normally scattered in the coffee trees with not many cherries per tree. This enabled pickers to move at a high speed.

Stages of coffee berry development.

Stage 1 Young Flower Bud

The first stage is a young flower bud. The tree branches produce clusters of flower buds

Photo courtesy of Thunguri Factory

Stages 2. Fully blossomed flower

Fifth to sixth day after blossoming the flowers wither and fall off at the ring tip of the young berry

Photo courtesy of Thunguri factory

Stage 3: Withered flowers fall on the ground exposing the ring tip on the young berries.

Pictures shows withered flowers fallen on the ground and young berries with tip rings.

Photo courtesy of Thunguri factory

Stage 4: Green berries forming.

Green cherries grow and form with the ring on the tip

Photo courtesy of Thunguri factory

Stage 5: Yellow beans

Green beans mature into yellow colour before they fully ripen to red cherries.

Photo courtesy of Thunguri factory

CHAPTER 10

COFFEE PROCESSING

PRIMARY PROCESSING

About 99% of Kenyan coffee is ***Wet Processed*** (Washed coffee) while about 1% is ***Dry Processed*** (unwashed coffees). Choice of processing methods depends on such factors as availability of water, the pattern in which the crop matures, the weather at the time of harvesting and the size of the farm. Other factors include availability of pulping and drying equipment, labour, and the market price of the coffee versus the cost of processing, and the variety of coffee.

Beside fermentation and washing the key difference between the Wet and Dry processing, methods is the wet processed coffee is dried as parchment without the coffee pulp while for Dry Processing, ripe coffee cherries are dried with the fruit pulp attached.

WET PROCESSING

Washed coffee preserves the intrinsic quality of the bean and gives a more homogenous coffee reflected in the appearance of both the green and final roasted bean however if not carried out correctly can be detrimental to quality.

Wet processed coffees at their best produce a balanced cup with incredibly good acidity of varying intensities, a medium body, with a fine, delicate, and complex flavour.

The cup for a Wet Processed Coffee is more delicate with better clarity compared to Naturals (Dry Processed coffee) which tend to highlight fruity notes and have a muted acidity. Wet Processing involves pulping and fermentation of parchment before drying.

Wet Processing Stages

Important stages in Wet Processing include cherry picking, cherry sorting, pulping, pre-grading, fermentation, under water soaking, final washing, and final grading

Cherry picking

There are basically three ways of picking coffee.

1) Selective handpicking mainly used by both Co-operative Society farmers who own small farms and Plantation Farmers. Here only the ripe cherries are picked repeatedly until picking is complete
2) Mechanical harvesting by large plantations in countries like Brazil. Mechanical harvesting is carried out when the farm has attained about 70 % cherry ripening.
3) The Stripping Method is also used in countries like Brazil where all the cherries are stripped from coffee tree without selection. Farmers wait until most of the cherries on the tree are red before stripping.

Kenyan coffee is handpicked. This is a success story especially for co-operatives societies which are characterized by small farms. Farming here is basically a domestic practice where the family is solely responsible for their coffee farms providing the required labour.

Clean containers are used when picking to avoid contamination while coffee pickers are not allowed to pick any fallen cherries. During picking the cherries are kept under cover in a shed to protect them from intense heat.

After picking, small scale farmers first sort cherries at home for any immature, dry and overripe cherries, twigs, and foreign matter before delivering to the wet mill factory normally located near the farms. At the factory, the factory manager inspects the coffee cherries for further hand sorting to separate any unripe cherries present. Cherries accepted for primary processing range from a slight green tinge at the stalk of cherry to an overall deep red. Drought effected, green cherry, overripe, rotten, dry or insect damaged cherries are not accepted

Cherry picking

Photo courtesy of Thunguri factory

No wonder most of the speciality coffees are produced by co-operatives through their small-scale farmers.

The ripe cherries are put together and are subjected to wet processing while the unripe cherries are sun dried and delivered as Mbuni (Dry Processed coffee). Plantation farms are constrained by labour and the volumes of coffee when it comes to sorting cherries. Factories (Pulping Stations) are collectively owned by farmers enabling them individually to reduce operational costs. They also act as accounting units along the marketing chain.

Cherries are delivered for processing on the day of picking as any delay makes pulping difficult and triggers the pre-fermentation process.

Cherry sorting at the factory

Photo courtesy of Thunguri factory

Pulping

Pulping is done preferably within 12 hours after picking. This involves removing of the cherry outer skin leaving the parchment as the only cover on the coffee bean. Pulping is enhanced by water flow to remove the coffee cherry skin.

Cherry pulping

Photo courtesy of Thunguri factory

Cherry skins are removed completely to avoid loss of parchment beans. The Pulping Machine is adjusted according to the size of the cherries to avoid nipping or crushing the beans. Nipping produces defective beans called "pulper" nipped beans which are discolored or chipped beans.

Parchment is graded mainly into three categories based on density namely parchment one (heavier parchment), parchment two (second heaviest parchment), parchment three and lights. Parchment grading facilitates uniform fermentation. Heavier and lighter parchment are fermented separately according to grade as it takes less time for lighter parchment to ferment.

Parchment three grade depends on the size of the crop, if small in quantity a separate fermentation is not justified so it is mixed with lights.

Any remaining beans not pulped are separated to avoid producing off flavours. Partially pulped beans are re-passed again. To avoid contamination, the receiving hopper discs, knives, sieves, graders, channels tanks and floors are cleaned before starting to pulp. This ensures there is no contamination of a new batch with remnants beans from the previous batch that have a potential of turning into major coffee defects. Left over beans in fermentation tanks develop into stinkers which could reduce the value of the coffee by up to 50 %.After pulping, fermentation is the next stage.

Fermentation

Fermentation involves the breakdown of the thick mucilage layer on the parchment into simple non-sticky substances which are easily washed off. Mucilage generally inhibits drying and attracts dirt/dust.

Research indicates fermentation improves acidity of the coffee therefore improving the flavour. Flavour rides on acidity, you cannot have coffee with good flavour without acidity.

Fermentation also improves the colour of the bean giving it a good appearance with a normally greyish green or bluish green colour. Under fermentation produces sour coarse liquor. Research has also shown that the amount of time and temperature under which coffee is fermented determines the flavour notes in coffee.

Parchment coffee with mucilage in a fermentation tank

Fermentation Tank

Photos courtesy of Thunguri factory

Fermentation tanks are preferably built in the ground and are usually painted with acid-resistant paint, preferably black for better heat retention and absorption. They are normally roofed to retain warmth and prevent entry of rainwater or direct sunshine. The tank floor is smooth and without any cracks which can harbour remnant beans that develop into stinkers.

Wooden or non-abrasive materials are used for stirring and ease of washing. The depth of the parchment in the fermentation tank is less than one meter. The tank is cleaned after fermentation to avoid contaminating the next batch.

Fermentation of coffee generally takes between 24-36 hours depending on factors like: temperature, (e.g., when it is warm fermentation takes a shorter period and vice versa) thickness of mucilage and concentration of enzymes. When it is cold enzymes are used to speed up fermentation. The parchment is fully submerged to allow an even distribution of the enzyme in the coffee mass. The enzyme powder is premixed with water in a container to dissolve to facilitate even distribution. This mix is then added to the tanks and thoroughly stirred.

To check if fermentation is complete, a traditional method is used by taking a handful of the wet parchment from the tanks, washing it in a bowl and feeling for any grittiness. Grittiness indicates fermentation is complete. If the parchment still feels sticky, then fermentation is incomplete.

Once all the mucilage is completely broken down the coffee is washed thoroughly by filling the tank with clean water and stirring it with paddles. The water is then drained off. Washing is done by pushing coffee parchment against flowing streams of water in water channels. This process is repeated until the coffee is clean.

Washing off mucilage from parchment coffee in Ethiopia. (Coffee is pushed against the water flow**)**

Soaking

Once all the mucilage in the tanks has been removed, the tanks are closed and filled with clean water to cover the parchment. The coffee parchment is left under water for between 16 to 24 hours. The soaking period can be extended so long as the water is changed daily. Salty water from boreholes is not good for soaking

In the event of an emergency, fully wet parchment is stored under water rather than in a wet heap. Soaking improves the cup quality and the colour and appearance of the bean by removing any brownish substances. It also loosens the silver skin for easy removal during polishing after milling. Once the soaking process is over, coffee parchment is washed out of the tanks for final washing and grading

Parchment drying

All Kenya coffee is mainly sun dried. Drying machines are used in large estates during peak harvesting. The drying process is done as quickly as possible to avoid mould developing. The moisture content of coffee parchment after washing is approximately 55 %.

Drying is carried out on raised drying tables made with fine mesh wire netting which allows upward draughts of warm air for faster drying. The depth of the parchment is about 2.5 to 5 centimetres to allow penetration of warm air and to avoid any dampness. The depth depends on the drying stage.

Estate farmers use mechanical drying when the weather is unfavourable. According to the "Coffee Grower's Handbook" published by the Kenya Coffee Research Institute there are five major drying stages.

Parchment coffee on drying tables with a hessian clothe in Ethiopia

Coffee parchment drying table with wire mesh underneath in Kenya

Photo courtesy of Thunguri factory

Parchment harvested at different times is dried separately. This ensures uniformity and avoids defects caused by uneven moisture distribution.

Drying stages

(a) ***Skin drying*:** (55-45% moisture content). This involves rapid removal of water from the surface and water lodged between parchment and the bean. Parchment is placed on wire meshes and is turned regularly to facilitate faster evaporation and exposure to sunlight.

(b) ***White Stage Drying**:* (44-33% moisture content)
At this stage the coffee bean is whitish; therefore, it is easy to sort out the defective beans. This stage is slower and controlled where parchment is shaded during the hottest part of the day. This helps avoid cracking of the parchment.

(c) ***Soft Black Stage**:* (32-22% moisture content) The final bean colour is formed at this stage. 48-50 hours of sunlight is recommended.

(d) ***Medium Black Stage:*** (21-16% Moisture content) The coffee bean is hard and black all over; it can be thoroughly dried without any effect on quality.

(e) ***Hard Black Stage**:* (15-12 % Moisture Content) A fully hard coffee bean, dark in colour is found at this stage.

(f) ***Fully dry and conditioning*:** Coffee is fully dry at this stage to 11.5 % moisture content. Small scale farmers use traditional methods to check whether coffee is dried to the right moisture content. These methods include, looking at the colour of the beans and squeezing the parchment coffee between their fingers. If the parchment and silver skin come off the bean, the correct moisture content has been attained or biting the coffee bean. If it breaks with a sound, then it's dry for storage. Another method used is rubbing parchment with fingers to feel the dryness. Portable Moisture meters and professional services from millers are also available.

Conditioning

Conditioning refers to storage of coffee after drying to even out moisture before delivery to the Dry Mill. Conditioning is done in ventilated stores or bins by blowing dry air to even out the moisture content of the parchment.

Dry Parchment Coffee

Coffee stores must be dry and leak-proof, burglar-proof, and fitted with racks about 3 inches from the floor. Storage of parchment over six months can result in bleaching and fading, a coffee defect. Coffee can acquire a woody flavour note at its extreme stage. Storing coffee parchment not dried properly also causes uneven coloured or faded beans.

Parchment with a high moisture content when stored for long periods can lead to mould growth, eventually resulting in black beans or rotten beans in its worst case. Black beans produce a musty flavour in the cup. Moisture meters help in monitoring moisture level. Coffee once conditioned is delivered to the mills immediately for security reasons.

Natural Processing (Dry processing) Picture of Drying cherries on raised tables

Dry Processing

Kenya Dry Processed ("traditional" Mbuni) Mbuni is a Kenyan type of unwashed (dry processed coffee) coffee. It includes unripe green cherries picked at the end of season, overripe cherries dried on trees and cherries fallen to the ground that are picked, sun dried and pulped. This is unlike specialty and non-specialty coffees that are deliberately dry (naturally) processed in other countries like Brazil. The moisture level of Mbuni is around 12 %.

The quality of Mbuni coffee is normally inconsistent in roast, full of softs and pales. It can be inconsistent in cup with a fruity, harsh cup with earthy, musty and occasionally fermentation notes present, due to the presence of a mixture of overripe and unripe cherries, unlike the "traditional natural" coffees from a country like Brazil.

Mbuni is graded into Mbuni heavy (MH) and Mbuni light (ML). Grading is by a gravity table and is based on bean density.Mbuni from Kisii County are better in quality with a more even roast and a consistent cup. Kisii grows the blue mountain variety.

Picture of Dry Processed coffee being spread on raised tables to dry.

Currently there are small quantities of Kenya dry processed ("Natural") coffees produced for the specialty market using the traditional "Natural" method. This method is more experimental and has not taken root in Kenya.

Picture of Thunguri coffee factory in Kenya dry processing their coffee.

Photo courtesy of Thunguri factory

The red ripe cherry is covered in almost airtight plastic bags to ferment for 2 to 3 days depending on desired flavour notes and then it is dried. Drying cherries rapidly with a high loss of moisture within a short time can affect the quality of coffee.

The Dry Processing (natural) Method involves drying the whole cherry. There are some variations between farmers on the process which has three main stages after picking the ripe cherries.

Stage one involves separating unripe, overripe, damaged cherries, twigs, and leaves.

In the case of mechanical harvesting, assessment is carried out before harvesting ensuring a higher percentage of ripe cherries.

Stage two is drying the cherries on raised African drying tables, large concrete, or brick patios. During drying the cherries are raked or turned to ensure even drying to 12.5 % maximum moisture content

Drying is the most important stage for this process due to its effect on the final quality. Over dried coffee will be brittle producing too many broken beans during hulling while under dried coffee with high moisture content is prone to rapid quality deterioration.

Stage three (hulling of the dried cherries)

This involves mechanical removal of the dry cherry pulp from the green bean ready for market.

Natural processing- (Ripe cherries spread on drying table to dry) in Ethiopia

Dry Process Methods in other Countries

There are various dry processing methods used in other coffee growing countries.

Pulped Natural (Honey processed or semi washed process)

Pulped Natural (Honey) Process

This involves pulping the ripe cherry (removal of the cherry flesh), drying the parchment with mucilage attached. Some Pulped Natural (Honey Processed) are processed and classified into three categories. Although there are inconsistencies in the actual processes from farm to farm, slight differences to the process are employed to improve quality and produce unique flavour notes. Although the name has nothing to do with the taste, honey processed coffees are known for their sweetness.

1)**White Honey** is mainly mechanically washed and has minimal mucilage left on the parchment before drying.

2)**Yellow Honey** is often semi washed with slightly more mucilage left on the parchment before drying.

3)**Gold, Red, and Black Honey** are differentiated by the amount of exposure to light and the drying time.

Gold Honey is dried during sunny times with less humidity which promotes quick drying.

Red Honey is processed under more shade to slow down drying time while at the same time increasing the amount of humidity the bean is exposed to.

Black Honey is dried under even more shade taking a longer time to dry than both gold and red honey coffees. These processes require more resources and if not carried out correctly can be detrimental to quality.

The main difference between the Honey and Natural Process is that pulping, or removal of the cherry fruit flesh is done before the drying, unlike natural processed coffee where the whole cherry is dried.

The Sumatra Wet Hulled Coffees Process

Indonesia has a unique processing method which is a hybrid of the Washed and Dry (Natural) process. This process is understood to significantly reduce acidity while increasing the body of the coffee, a unique attribute of the Sumatran coffee cup profile. It also gives the coffee some earthy and winey notes.

After pulping, the parchment is fermented overnight in a concrete tank or plastic bags to break down the mucilage. This is followed by washing the following day. Wet parchment is dried to a moisture level of 20 % to 25 % for hulling (removal of parchment).

The moisture level may vary from one farmer to another, but the bottom line is that the Sumatran coffee process dries the coffee parchment halfway, before hulling, prior to the final drying stage where the bean is dried to a moisture level of about 12 %, suitable for storage.

The final drying stage gives the bean a unique deep olive greenish colour. Its cup has a complex earthy flavour, low acidity, and heavy body. While other countries may view this method as defective, Sumatran coffees have their own markets where they are used in blends.

Sumatran Green Coffees are hand sorted into three categories, Single Picked, Double Picked or Triple Picked lots. Hand sorting does not necessary improve the cup but does improve the appearance of the bean.

It is quite easy to over roast Sumatran coffees because the subsequent colour development during roasting is influenced by this unique drying process. Sumatra is an island in Indonesia while Mandheling is associated with an ethnic group from that island. Sumatra Mandheling, one of the main types of Sumatran coffees is named after this ethnic group while the other coffees are named after their growing regions e.g., Sulawesi

The Mandheling description is said to have originated from a Japanese coffee buyer who returned to Medan in 1969 seeking a coffee that he had tasted during his WWII service. It is said during his service he asked a local where his coffee originated from. The local answered Mandheling thinking he was being asked about his ethnicity.

Sumatra Grade 1

Kenya AA grade

The main Stages for different Primary Processing Methods

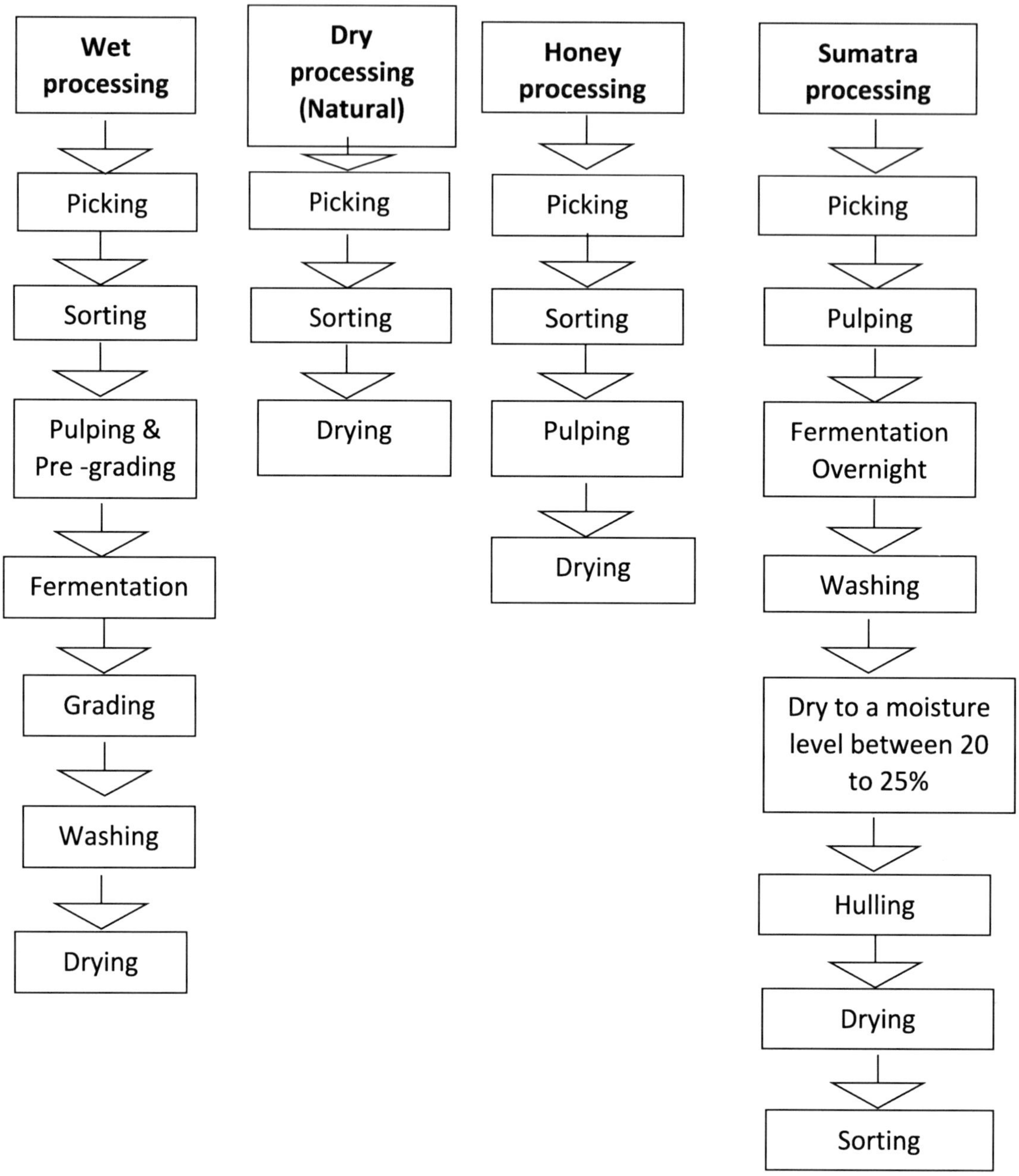

CHAPTER 11

SECONDARY PROCESSING (Dry milling)

Dry milling machinery

Photo courtesy of Nairobi coffee Exchange

Milling is a crucial process in coffee quality and the value adding chain. In Kenya we have both commercial and private mills. Milling is the process where the parchment and dry fruit pulp covering of the coffee bean is removed. The milling process results in a weight loss, termed the milling loss. The primary process used is the main factor in determining the milling loss.

Other factors include the density of the bean. Dry processed Mbuni (MH) coffee can lose up to 50 % in weight due to the removal of the pulp. The milling loss for wet processed coffee is much less as it is restricted to loss of the parchment.

Dry Milling Stages

1) Initial Analysis of parchment and dried cherry.

Once coffee is received in the warehouse, a sample is drawn, and its moisture is analysed to determine whether it requires further drying before milling. Laboratory analysis is carried out to estimate the weight loss after milling and the resultant coffee bean is cupped. This quality analysis enables bulking of similar small coffee lots.

2) Pre-Cleaning

Before commencing milling, the parchment or dried cherries are pre-cleaned. This involves removal of debris, stones, dry pulp, metallic parts, dust, and sisal fibres. These objects are removed using gravity separators, sieves, and magnets.

3) Hulling (Removal of pulp and parchment.)

The coffee goes through the Huller where the parchment or dried pulp is removed. Hulling of coffee with parchment only and coffee with dried pulp is carried out separately.

4) Polishing

Polishing is carried out after hulling to remove part of the Silver Skin. Coffee beans when put under a lot of pressure (over polishing and total loss of Silver Skin) can develop a brownish colour, detrimentally affecting bean appearance and shortening of storage life.

5) Grading

Wet Processed Coffee

After hulling and polishing, Wet Processed coffee goes through the grading system where six main grades emerge, namely: AA, AB, PB, C, E and T. Lighter (less dense) beans from grades AA and AB are separated using a Catador (similar to a Gravity Table) producing another grade called TT. The lighter coffee from C grade is mixed with the T grade.

Dry Processed Coffee (Mbuni)

Mbuni is graded into AA, AB, PB and C grades using screens. The coffee is then passed over a gravity table and the lights (less dense beans) from each grade are bulked together with T grade which produce ML (Mbuni light) grade. The remaining coffee from the separation process of all the grades is bulked together producing MH grade (Mbuni heavy). The coffee then is bagged unless it is being passed through the colour sorter to remove defects.

Coffee grades, Description and Screen size

The grade retention on the screens specified should be about 95 %. It is never 100 percent and varies depending on the efficiency of the process

E Grade also known as "Elephant", is one bean made of two beans joined together, it is a genetic defective grade. E is normally mixed with some PB grade beans. The bean splits open into ears during milling, grading, and roasting causing a defect called "shells", The same also happens when roasted. It is retained on screen No. 21 (8.3 mm)

PB is a grade where one ovule develops in a cherry instead of the normal two. It passes through screen 17 and is retained on slotted screen No.12 (4.76 mm)

T grade consists mainly of broken beans and smaller grade C beans. It is separated from PB and bold C grades. It is retained on screen No.7 (2.9 mm)

TT grade consists of light density beans extracted from AA and AB grades by air using Catador. It is a mixture of screen sizes.

AA grade is the second largest bean after E grade. It passes through screen 21 and is retained on screen No. 18 (7.2mm).

AB is a combination of A and B grade. Previously these grades were marketed individually. It passes through screen 18 and is retained on screen No. 16 (6.35mm)

C grade is a small size bean. It passes through screen 16 and is retained on screen No. 10 (3.96mm)

Kenya Coffee Main grades and Screen size grading flow chart

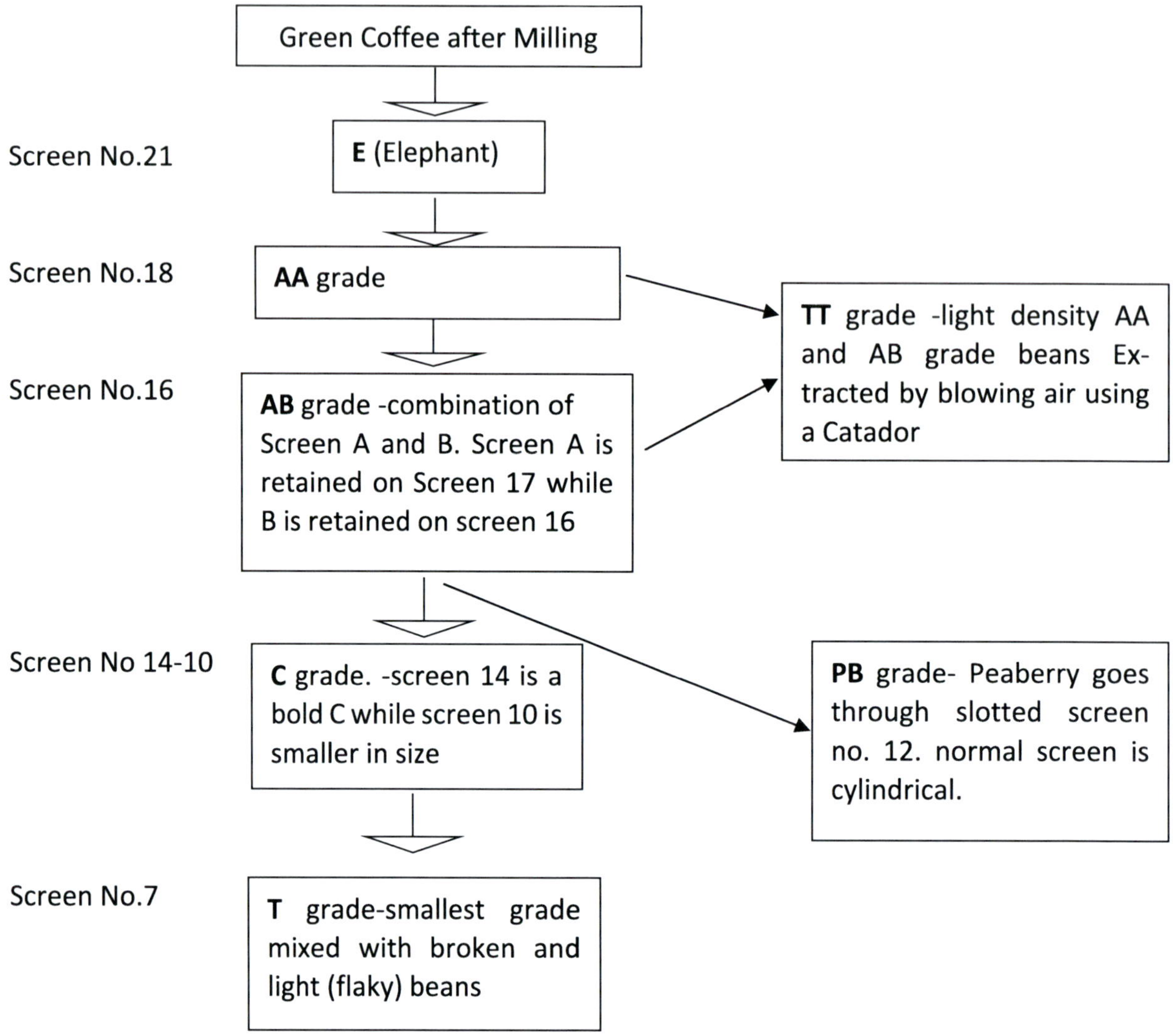

Slotted Screen used for grading PB grade **Cylindrical Screen for normal grades**

Other minor grades include UG1, UG2 and UG3 (ungraded coffee grades 1,2 and 3).

"Parchment Three" and "Lights" are milled and graded into UG1 (Ungraded Coffee grade I) and UG2 (Ungraded Coffee grade 2) or UG3 (Ungraded coffee grade 3). Grading is done using gravity tables and is based on weight, with the heavier beans being classified as UG1 and less heavier beans UG 2.

MH (Mbuni heavy) and ML (Mbuni light) are graded from unwashed coffee only.

It is important to note that lighter parchment (UG1, UG2 and UG3) when put through standard grading machines may produce deformed coffee beans, lighter in density and rugged or shrivelled in terms of bean formation.

Originally all parchment coffee was milled by the Kenya Planters Cooperative Union Mill in Nairobi located next to the Coffee Board of Kenya (currently Kenya Coffee Directorate), the only marketing agent at the time. The two buildings were interlinked enabling coffee to travel through a pipe once milling was complete to the Coffee Board warehouses for bagging, storage, and marketing.

Gravity Separator.

Photo courtesy of Nairobi Coffee Exchange

Sensory Evaluation.

After milling coffee is then cupped and evaluated for sensory properties. The Millers send a "Coffee Classification Report" to the farmers to help them identify the various faults and appropriate suggested corrective measures. The report includes for each given consignment the moisture content, quantity per grade, awarded class score, cup quality and percentage weight loss after milling

CHAPTER 12

COFFEE CLASSIFICATION AND QUALITY

Kenya Coffee classification standards

Grading of Kenyan Coffee is based on the physical screen size of the bean whereas coffee classification is based on quality and sensory properties including grade. Classification takes into consideration green and roasted bean appearance, cup quality, category of defects and number of defects.

Kenyan coffee is classified into 10 classes or standards of coffee. Class one being the best. For example, you can have AA grade class three and AA grade class six. The same consignment (Outturn) of coffee may include different grades with each grade having a different classification score. For example, two grades from the same outturn AA would score class 5 while AB scores class 6.

The Coffee Catalogue used for listing coffees for auction consists of a Main and Miscellaneous Section. The Main Section comprises Classes one to six while the Miscellaneous Section includes all other classes

The Kenya Coffee Classification System has no stated specified defect count for each class however it is used as a parameter to score the coffee, unlike other countries. Most of the coffee exported in the speciality categories is handpicked to remove almost all defects. Sours and blacks are not allowed in coffees offered in the "Main Section of the Catalogue".

Coffee with fermentation cup defects may be found in some grinder quality coffee and coffee destined for Middle Eastern markets where they consume their coffee with spices. These coffees would be listed in the Miscellaneous Section of the Catalogue.

Every coffee stakeholder must possess a license issued by the Kenyan Coffee Directorate who is responsible for the oversight of the Coffee Classification System.

Other Coffee Classification Systems

Several coffee growing countries base their coffee classification on the number of defects. Ethiopian coffee grades are determined by cup, physical appearance, and number of defects.

All Ethiopian coffee is retained on screen 14, and there is an emphasis on defect count and the geographic region where the coffee is grown.

Brazil uses specific names like Cerrado and Santos and screen size to identify the specific type or quality of coffee.

Countries like Guatemala use growing regions like Antigua and the height at which coffee is grown to identify their specific coffees. For example, "strictly hard bean" is coffee grown between 1500 to 1700 meters above sea level.

Quality Parameters used for evaluating Kenyan Green coffee beans.

Uniform bean with an Ideal colour

Size and uniformity of Green Beans:

Screen size varies according to weather conditions, variety, and region. The actual size of the bean other than uniformity does not have a significant effect on quality. Some of the finest coffee Kenya produces has smaller but even and solid beans.

Green Bean appearance: a white straight centre (white line) cut running through the flat side of the bean is an indicator of quality.

Green bean with brownish colour

Colour: This is of great importance with the ideal colour farmers should target being a bluish green or greyish green. Coffee of this colour has some of the finest flavour attributes. A brownish bean colour is a fault which can be avoided. Mechanical drying can also give a brownish colour.

A brownish colour can also be caused by insufficient pre-washing, drought, excessive rain, or persistent cold weather, overripe coffee and over fermentation with too many skins.

Defects: The number of defects and whether they are major or minor, is another factor that is considered when analysing quality. Major defects include sours and blacks while minor defects include brokens.

Roasted Bean Appearance- This refers to the general appearance of the roasted beans.

Centre cut- This is the dividing line covered with a silver skin running through the centre of the flat side of the coffee bean. After roasting, this centre-cut in a fine quality coffee stands out clean and white.

Evenness: - Uniformity of roast colour is a good indicator of bean quality.

Unevenness of roast bean appearance can result from insufficient separation of lighter coffees in the washing channel, uneven fermentation, or different picking or bulking of coffee from different regions.

Good quality even roast

Dull Poor-quality uneven roast

Cup

Cup quality is based on three attributes, namely acidity, body, and flavour. Cup defects like fermentation, foul, unclean and earthy off flavours lead to marking down the cup quality and coffee in general. A good cup is balanced in terms of acidity, body with complex flavour notes. Fermentation at its extreme can lower value of coffee by up to 50 %.

Sensory Uniqueness of Kenyan Coffee

Kenya coffee is known for its unique and complex flavour profile with high citrusy and lime acidity, floral, honey and raw sugar notes, giving it a unique place among other world coffees. A key attribute of the Kenyan coffee is its acidity making it appropriate for brightening the flavour of coffee blends. The popularity and demand for Kenyan coffee allows it to enjoy a price premium over other world coffees.

FACTORS THAT CONTRIBUTE TO COFFEE DEFCTS

Coffee quality is subject to several factors including the coffee variety, topographical and prevailing weather conditions during the growing season, husbandry practices, harvesting, storage and shipment conditions and transportation.

Variety and topographical conditions are constant; weather conditions may vary altering the quality from season to season with no human intervention. Crop husbandry, harvesting, processing, storage, shipment conditions and transportation are variables that are subject to human influence.

Coffee defects whether minor e.g., pulper nipped beans, ears, and broken beans, or major e.g. stinker and blacks are a measure of coffee quality limitations caused by deficiencies in husbandry and processing practices or adverse climatic conditions.

A quality controller cannot reverse primary quality faults but can only identify the cause of a quality problem, note, and point them, but have no authority or means to enforce preventive measures.

Effects on quality occurs at two stages.

(a) Husbandry (Inherent quality of coffee).

(b) Harvesting, and post-harvest.

(a) Husbandry of Coffee

Farm Husbandry determines the quality of the coffee bean produced. It determines bean characteristics like size, density, and the bean's formation. Factors including drought, lack of fertilizer or manure, not spraying or pruning and lack of well distributed rainfall in general negatively impact quality during this development stage.

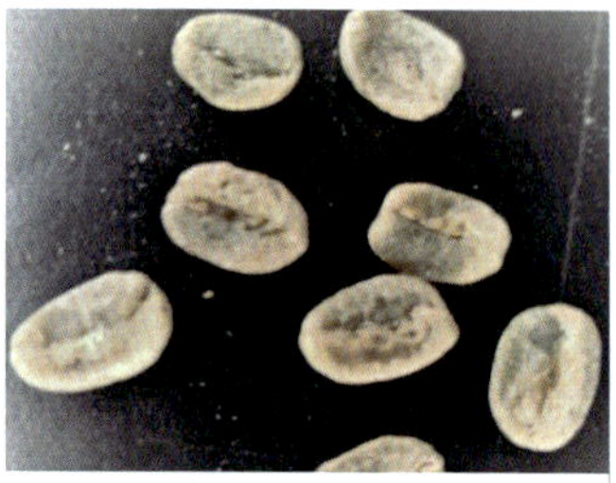

Drought Affected

(b) Harvesting and post-harvest

Harvesting methods and poor processing procedures can cause the inherent quality of the bean to deteriorate. Handpicking of coffee cherries, a common method of harvesting coffee in Kenya involves selective picking of the red cherries. While this method is labour intensive it optimizes bean quality. Harvesting of overripe beans results in foxy beans. Overripe cherries (foxy beans) produce a fruity to sour cup.

Stripping is another method of harvesting which is mainly used in large estates (plantations). It involves stripping the coffee branches of both yellow and red cherries. Stripping is organised at a time when there is a small percentage of unripe beans.

Foxy Beans

Machine harvesting is used in countries like Brazil with large coffee plantations that have a more homogenous maturation of cherries at the time of harvesting.

Green and red Cherries during picking

Small farms handpick their coffee due to the land topography. Technology is used to sort out the mechanically harvested cherries, separating overripe and under ripe cherries to improve quality. Harvesting of green cherries together with ripe cherries often gives the cup an unpleasant greenish, hay taste that diminishes with time. It can also affect the uniformity of the green and roasted beans appearance. Immatures beans manifest in the roasted beans as softs and pales (Quakers).

Pales/Quakers

Pulping of coffee cherries requires plenty of water to remove the pulp as the cherries pass through the hoppers. If not, properly adjusted the pulping machine allows some whole cherries to pass through un-pulped. It could also chip some beans producing discoloured and partly chipped beans, a defect called pulper nipped beans.

Pulper Nipped

Failure to pulp cherries immediately or a few hours after picking triggers a pre-fermentation process which impacts on the quality. This can lead to stinkers or sours. Stinkers smell when crushed and produce a fermentation "note" in the cup. This can reduce the value of the coffee by up to 50 % of the original value.

The Fermentation Process is an important part of wet processing, however, over fermentation of coffee can do more harm than good. It can produce a very fruity, fermented, or foul cup in its extreme (Stinkers or Sours). Clean washing water must be used in wet processing. Using polluted water produces an "onion" flavour or unclean cup.

Stinker/Sours

Coffee Washing

Photo courtesy of Thunguri factory

Wetting of already dry parchment or green beans during transportation, loading, offloading or in case of a leaking roof during storage can ruin quality. This leads to development of water damaged beans. As a defect it is recognized by deep greenish tips or patches. Green coffee when water damaged should not be dried directly under the sun. Excessive moisture can result in developing moulds and rotten beans.

Milling process:

Wet processed parchment in Kenya is dried to 11.5% moisture content while Dry processed Mbuni is dried to 12 % moisture content before milling. This level of moisture not only preserves quality after milling but ensures the bean is hard and cannot be easily damaged during milling.

On the other hand, coffee with moisture lower than 8 % loses most of its flavour. Milling faults can result in minor defects namely ears and broken beans. These defects do not affect the cup but have a detrimental effect on roasted bean appearance. Ears break away from E grade.

Ear/Shells

Broken Beans

Milling standards associated with a grading system over the years has ensured maintenance of the quality of Kenyan coffee. Milling can affect the quality of coffee through discolouring of beans (e.g., during polishing).

Milling of coffee with high moisture leads to quality deterioration shortening shelf life. Millers ensure coffee accepted for milling has the right moisture content and if need be, coffee is dried first before milling and the farmer charged. This helps avoid fading and rapid ageing of coffee.

Faded Beans

Packaging

Packaging for storage and export can impact on the shelf life and inherent quality of green coffee beans. Research has established the detrimental effect oxygen, humidity and moisture can have on coffee over time. Quality deterioration in aroma and coffee flavour can become evident after about six months of storage.

Vacuum packaging is mainly used for speciality Kenyan coffees to preserve quality. It is however expensive compared to other alternatives including grain pro bags, (jute bags with a sealed plastic bag inside).

Storage Conditions

The storage of parchment (green bean with husk) or green coffee is important when it comes to quality. The husk and silver coating on parchment coffee can serve as a barrier to external damage during storage. Parchment coffee once dried should be placed on wooden floors or ventilated bins. Parchment is normally ready for delivery to the mills within a month after it is fully dried. It should be stored in an environment where it cannot gain or lose a lot of moisture to avoid spoilage. High humidity and high temperature accelerate fading of the coffee beans.

A coffee store should not store chemicals or any other substance of this nature as coffee will pick up these chemical aromas.

Coffee for export should not be stored for long in coastal warehouses where most ports are located due to the high humidity and temperatures of these coastal regions. Aged coffee at its extreme can develop a woody flavour in the cup.

Storage Time:

The longer coffee is held in storage before milling, after milling or before consumption, the more it deteriorates in quality. Storage for a long time can lead to weight loss due to loss of moisture.

CHAPTER 14

THE KENYAN AUCTION SYSTEM

History

Between 1930 and 1933 coffee marketing was handled by individuals. Prior to the establishment of the auction system in Kenya, farmers shipped parchment coffee for milling and marketing to Britain, London Mincing Lane.

This was a difficult time for farmers who had to wait for about six months for payment from their traders. The other challenge that faced farmers was that only the best quality was sold while rest of the coffee remained unsold in the British warehouse. The first auction was held in July 1931; however, farmers took a long time to accept the system.

In 1936 the East African Coffee Exchange was established incorporating Uganda and Tanzania, with auctions being held in Nairobi and Mombasa.

In 1937 the Kenya Coffee Board took over the Kenya Coffee Auction (KCA) having formed it in 1934. The year 1963 saw the end of the East African Coffee Exchange with every country independently managing their own industry.

Coffee Auction Hall

photo courtesy of Nairobi coffee exchange

1st October 1997, the Coffee Board introduced multi-auctioneering of coffee at the Nairobi Coffee Exchange. This was meant to offer farmers alternative auctioneers. Before then all coffee was auctioned through the Kenya Coffee Auction the only auctioneer at the time. The push to introduce more auctioneers came from farmer's representatives.

With the introduction of marketing agents, the auctioneer's role of working with the Coffee Board shifted to working with the different marketing agents. Auctioneers carry out actual selling of coffee on the floor of the Nairobi Coffee Exchange on behalf of the marketing agents.

The Current Kenya Coffee Auction System.

Nairobi Coffee Exchange New auction system display board. It displays catalogue sale number (25), the lot number (1280) being offered Societies name and reference (KIANGAI/XDA07F04), Mill reference(23NG0063) and grade of coffee (ML), Number of bags (21 and weight is 1319 kilograms), Certification (FLO)

The Kenya Coffee Auction is hosted by the Nairobi Coffee Exchange. Auctions are held once every week on Tuesdays apart from public holidays, depending on availability of coffee. Licensed coffee dealers (Buyers) are the only entities allowed to bid at the auctions with the highest bidder above the reserve price buying a given lot.

The Coffee Catalogue with the various lots of coffees being offered in every auction is prepared offering various Classes (qualities) of coffee. The Main Section of the catalogue offers top and commercial quality coffees, while the Miscellaneous Section of the catalogue offers lower quality and defective coffees.

The Catalogue allows buyers to purchase different qualities according to their different market requirements.

Samples for all the lots offered in every given auction sale are delivered to the Trade Sample Room for collection by the Coffee Dealers' Representative. Buyers then evaluate them ready for the auction. Samples are availed for collection by buyers ten days before the day of the auction date.

Auction Trade Sample room (Pre-offer Auction Samples)

Samples from different lots on offer for an auction sale are arranged in the Central (Trade Sample Room) for collection by buyers before the auction.

Coffee Catalogue

As earlier mentioned, the catalogue is normally divided into two sections:

(a) ***The Main Section***: This part of the catalogue consists of commercial, premium, and top qualities and includes predominantly main grades.

(b)***The Miscellaneous Section:*** This part of the catalogue consists of coffee with major defects, the ungraded coffees, and the unwashed coffees.

This distinction is deliberate and is supposed to indemnify the seller of the coffee against unwarranted quality claims arising from quality defects and faults. Buyers are fully aware of the quality in each section.

The auction system enables the dealer to access coffee of all qualities from both co-operative and plantation (estate) sectors from a centralised source.

The Catalogue shows the offer sale number, auction venue, date and time of the auction, terms and conditions of sale, and the Prompt Date, which is the date by which the payment of coffee bought at an auction, must be settled. It also details the coffee lot number, outturn number (identification number allocated to every consignment delivered for milling and marketing), marks (name of the farm or cooperative society), the number of bags offered for that a particular lot and corresponding weights.

Once the coffee is tasted, the cupper makes notes against every lot number tasted and evaluated, establishing its value and possible usage based on its quality. Notes are meant for quick reference during biding so they should be precise and definitive.

From the onset the Coffee Auction was conducted through the 'hammer system', where bids were offered by word of mouth and a show of hands and concluded by the hammer.

Since April 1998 the coffee board has introduced a computerised trading system. The modernized computerised trading system at the Nairobi Coffee Exchange has among its main features an **Auctioneer Terminal** which is used to run the auction by the auctioneer.

It also has a **Seat/Key Unit,** software used to allocate seats to the dealers (buyers) either on a random basis or on a pre-determined manner including programming the dealers' keys.

A **Display Board** at the front of the auction hall enables the auctioneer and the dealer to communicate as the bidding is going on.

A key feature is the **Push Button** on every dealer's seat for bidding. A glowing light simultaneously indicates the winning bidder both on his seat and the display board. The buyer's name is also displayed on the display board and the price at which he has bought a lot of coffee.

Computerisation of the auction system has enhanced transparency and integrity of the bidding process besides being more efficient. Currently the auction system has been upgraded allowing buyers to bid remotely (from their offices).

The auction system guarantees the farmer the highest prices he can obtain in the local market based on prevailing world prices. The system acts as a platform for fair and transparent competition driven only by quality, demand, and supply. The auction exposes every lot of coffee to the forces of demand and supply at the floor of the exchange.

The Auction Process

Buyer's preparation for the Auction

The Buyer's cupping room is normally a beehive of activity as they prepare for auction every week. Once the samples are collected, quality analysis which includes assessing defects, coffee condition, screen size and cupping commences.

Two cups are normally tasted for every lot in the Main Catalogue to ensure consistency in quality judgment and absence of taste defects. Three to five cups for miscellaneous lots are tasted due to the high risk of defects being present. The number of cups tasted can also depend on the buyer's quality regime.

Numbers of overall samples offered every week could range between 500 to 700 samples. The number of bags on offer for a given sale week is dependent on expected yearly production, prevailing prices, and price trend.

Sample Preparation

Once the samples have been collected, they are then put into trays labelled sequentially ready for roasting. Roasting is a key part of the sample evaluation and quality decision making process. Roasting drums are emptied to avoid any contamination between lots.

Samples are roasted to a medium to light roast to bring out cup attributes or any off flavours. Over roasted coffee supresses and masks both cup sensory attributes and defects.

Sample contamination during this process with defective coffee can lead to forfeiting a bidding opportunity against competitors. If unsure of the sample cup attributes, it is always advisable to re-roast the sample.

Grinding is another key aspect to the evaluation process. Grinding is carried out systematically to ensure there is no sample mix up or contamination. Getting the sample preparation process right ensures the buyer can be confident of quality decisions and notes made are accurate, as they spend thousands of dollars based on these notes. Once the first round of tasting is complete, the buyer would re-taste (commonly referred to as re-cup in the industry) certain samples, for validation.

Re-tasting is also done for higher qualities to reaffirm quality scores as high prices are paid for this coffee. The other reason for re-tasting is when there is significant disparity in quality scores between coffee sister lots (these are different grades of coffee of the same outturn offered as individual lots within the same catalogue). For example, where the AA grade cupping score is quite different from the AB grade from the same outturn in the same catalogue or one lot is clean, and the other is not, or is fermented.

Once the quality analysis is complete at the respective buyers' cupping room, final visual verification is carried out on the green bean samples held at the Trading (central) Sample Room. This re-affirms that samples collected and tasted are the correct samples. If there is any doubt, fresh respective samples are drawn and tasted again. If there is inconsistency in quality assessment on these redrawn samples, the lot is withdrawn.

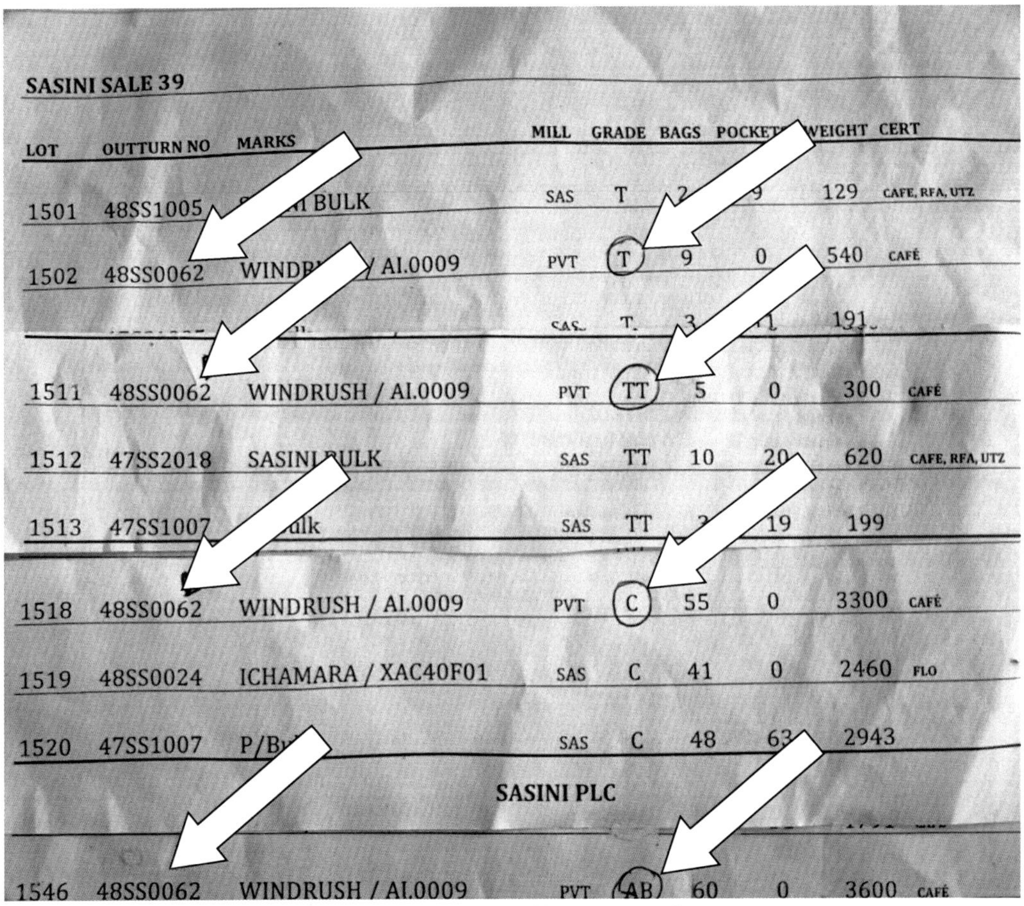

SASINI SALE 39

LOT	OUTTURN NO	MARKS	MILL	GRADE	BAGS	POCKET[illegible]	WEIGHT	CERT
1501	48SS1005	[illegible] BULK	SAS	T	2	9	129	CAFE, RFA, UTZ
1502	48SS0062	WIND[illegible] / AI.0009	PVT	T	9	0	540	CAFÉ
[illegible]	[illegible]	[illegible]	SAS	T	3	1	191	
1511	48SS0062	WINDRUSH / AI.0009	PVT	TT	5	0	300	CAFÉ
1512	47SS2018	SASINI [illegible]ULK	SAS	TT	10	20	620	CAFE, RFA, UTZ
1513	47SS1007	[illegible]ulk	SAS	TT	[illegible]	19	199	
1518	48SS0062	WINDRUSH / AI.0009	PVT	C	55	0	3300	CAFÉ
1519	48SS0024	ICHAMARA / XAC40F01	SAS	C	41	0	2460	FLO
1520	47SS1007	P/B[illegible]	SAS	C	48	63	2943	

SASINI PLC

LOT	OUTTURN NO	MARKS	MILL	GRADE	BAGS	POCKET	WEIGHT	CERT
1546	48SS0062	WINDRUSH / AI.0009	PVT	AB	60	0	3600	CAFÉ

Catalogue sister lots. Same outturn but different grades

Auction Samples Evaluation

Coffee quality is mainly determined by the cup hence tasting the cup for the correct lot sample is paramount. Correct identification of the samples is very important.

Sampling

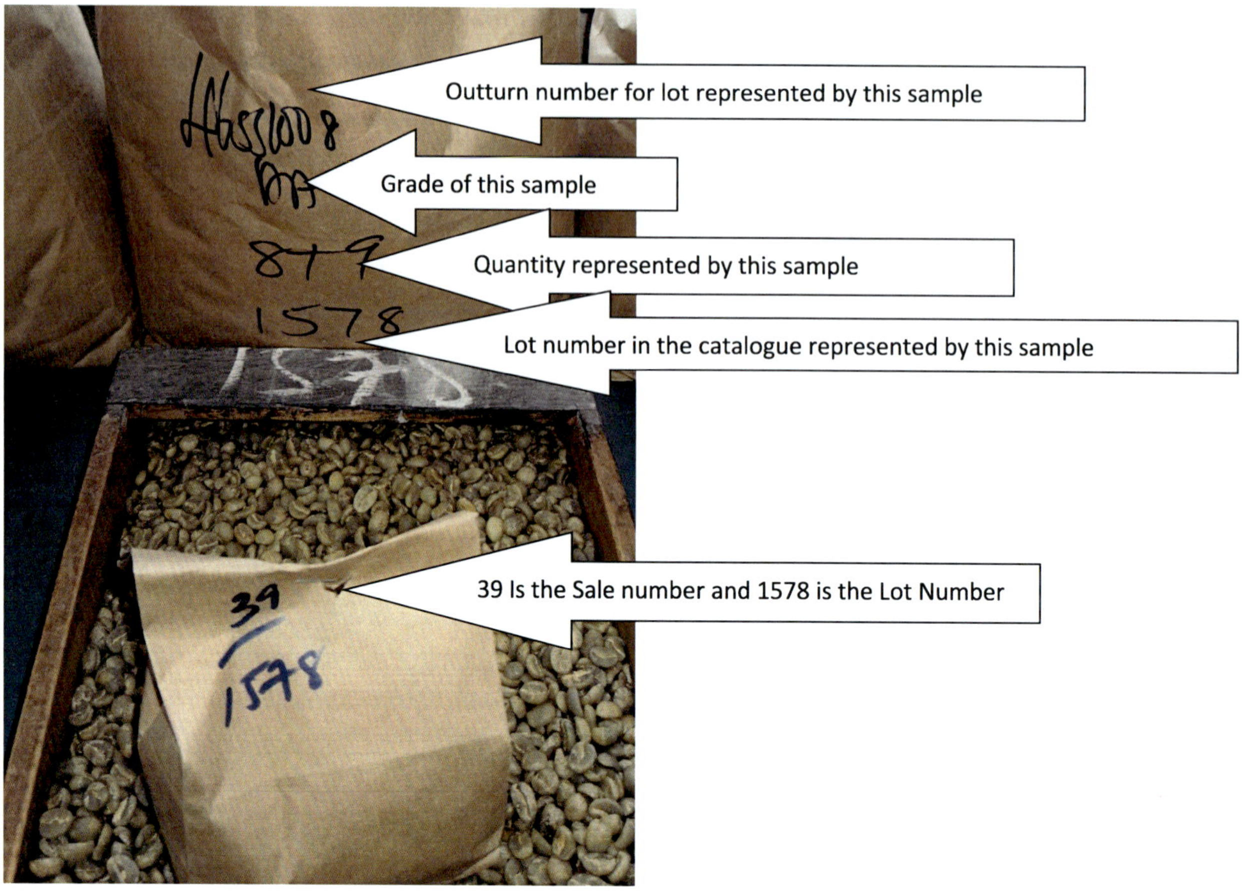

The Sample is identified with the outturn number, the grade and lot number on offer in each catalogue.

Auction Buying

Once an exporter is ready for the auction, the next step is actual buying of coffee at the auction. The auction is carried out in an auction room which is open to buyers, farmers, and members of the public.

There is now provision for conducting the auction remotely. The Auctioneer offers to sell coffee at the auction on behalf of the farmer. The buyer or the buyer's agent must be registered with the Nairobi Coffee Exchange. The buying process is conducted lot by lot. The speed at which the process moves requires quick decision making and alertness by buyers, who assume ownership of the coffee once the light on the push button turns red.

Trading Terms of the Kenya Coffee Auction.

The Prompt Date is seven days from the date of sale when payment must be settled. Payment delay by buyers attracts interest and storage charges.

10 Societies and estates coffees that fetched the highest prices at Auction Over 10 years. Ranked from 1 to 10 (highest price to the lowest price).

	2009/2010	2010/2011	2011/2012	2012/2013	2013/2014	2014/2015	2015/2016	2016/2017	2017/2018	2018/2019
1	Gathiruini	Kianjiru	Khachonge	Mukui	Iyego	Kiriani	Barichu	Ndaro-ini	Karimikui	Gititu
2	Kianjiru	Gathiruni	Chepkube	Kiringa	Kiangombe	Gikirima	Kamwangi	Tegu	Aguthi	Kiangoi
3	Karimikui	Karimikui	Kibisi	Tambaya	Konyu	Kamwangi	Tegu	Kangocho	Gakuyu-ini	Gatomboya
4	Kiangoi	Kiangoi	Kithino	Karumandi	Karimikui	Kiangoi	Gatomoboya	Gatomboya	Karimikui	Warenew
5	Kii	Kii	Kerumbe-Estate	Karimikui	Gachami	Thithi	Karimikui	Karimikui	Kii	Chesiro
6	Kianjogu	Kirimara	Karimikui	Kii	Karumandi	Gachami	Kii	Kiangoi	Ruthagati	Karimikui
7	Gachami	Kianjogu	Kii	Kiangoi	Kamagogo	Kii	Gititu	Kii	Kamwangi	Githaka
8	Ndimani	Ruthagati	Karumandi	Kianderi	Guama	Kagongo	Gaturiri	Karatina	Gititu	Marua
9	Kaliluni	Oaklands-Estate	Doondu-Estate	Ruthagati	Gondo	Gakuyuni	Ndima-ini	Kianjiru	Ndaro-ini	Kieni
10	Twin River-Estate	Gatomboya	Kapkwen	Guama	Challem-Estate	Jungle-Estate	Karindundu	Gachatha-ini	Iyego	Kii

Source: Nairobi Coffee Exchange

Auction Prices

Marketing agents are responsible in consultation with farmers for setting reserve prices for every lot of coffee offered based on both the prevailing New York C Market and the local market.

The actual price for every lot is determined on the auction floor through competitive bidding based on quality, demand, and supply and the prevailing New York C Market.

The price is also influenced by speculation with buyers anticipating price movements upward or downward, therefore buying more than their commitment or short buying.

Uncertainty regarding overall stock availability may also influence prices. The auction system ensures price discrimination between the different qualities in each auction with better quality coffees fetching high prices and vice versa.

Variation in coffee supply from Brazil, a major producer of world coffee affect coffee prices. Brazil has been the world's largest producer of coffee for more than 150 years (Carley Clontz October 1,2018) supplying about 30 % of world coffee.

The Kenya Coffee Auction has earned itself a name over the years in the world of coffee. Biding at the auction is purely based on the samples provided to the buyers. Buyers spend millions of dollars on lots of coffees offered based on these samples. The Auction is accessed by all stakeholders. Premiums for certified coffee are payable only for coffees that are declared and sold as certified.

The role of the Kenya Coffee Board now known as Coffee Directorate has been scaled down to regulating the coffee industry, which includes, implementing government policies and ensuring that the industry players operate within the set rules of the industry.

20 BEST FACTORIES PAYING RATES IN 2017/18

County	Co-operative	Cherry payment rate per kilogram (In Kenya Shillings)	Factory	Cherry (Fresh picked red cherry) Production in Kilograms
Nyeri	Gachatha	105.00	Gachatha	427,696.60
Kirinyaga	Thirikwa	103.00	Gakuyu-ini	1,102,119.00
Kirinyaga	Rung'eto	100.20	Kii	783,449.00
Nyeri	Gikanda	98.94	Ndaroni	419,929.00
EMBU	Kibugu	97.45	Gicherori	460,591.00
Kirinyaga	New Ngariama	97.05	Kamwangi	781,900.00
Kirinyaga	Karithathi	97.05	Kabingara	508,486.00
Kirinyaga	Baragwi	96.95	Kagongo	636,322.00
Kirinyaga	New Ngariama	96.50	Kainamui	611,406.00
Kirinyaga	Karithathi	96.20	Kiunyu	924,145.00
Nyeri	Mugaga	95.75	Kieni	557,637.00
Nyeri	Othaya	95.50	Karuthi	138,551.00
Kirinyaga	Rung'eto	95.20	Karimukui	1,262,422.00
Kirinyaga	Baragwi	94.82	Guama	1,488,487.00
Kirinyaga	Baragwi	94.76	Thimu	1,410,825.00
Kirinyaga	Baragwi	94.25	Mushagara	1,008,180.00
Kirinyaga	Baragwi	94.23	Karumandi	1,306,354.00
Kirinyaga	Baragwi	94.18	Gachami	990,056.00
Kirinyaga	Baragwi	93.61	Kariru	675,268.00
Kirinyaga	Baragwi	93.55	Nyanja	745,108.00

Table shows 20 factories with best paying rates for cherries in 2017/18.

Settling Quality Claims

The auction system has a quality claim resolving mechanism built into it. Every lot of coffee offered at the auction has three samples drawn from it.

The first sample is the Pre-offer Sample (Pre-auction sample) which is collected by the buyer before the auction for their analysis. The second sample is the Buying Sample which is given to the buyer for every lot of coffee they buy after the auction. The Buying Sample is compared with the Pre-offer Sample immediately after the auction for validation. The third sample is a Reference Sample that is kept under the custody of the Coffee Directorate for six months. If the Pre-offer Sample does not match the Buying Sample, another sample is drawn.

If the Re-drawn Sample is different from the Pre-offer Sample the buyer registers a claim in writing with the Auctioneer. The claim is supported by the Pre-offer Sample and Buying Sample which are then compared to the Reference Sample. If there is disparity the buyer is then issued with a credit note equivalent to the amount paid, and the coffee is re-offered at the next auction with the correct Pre-offer Sample.

Weight Claims

Coffee is sold at the auction as per the weight on the Warrant which is the Coffee Title Document. Warrants show the gross weight, tare weight and moisture content in addition to other details.

Short weight claims are lodged with the warehouseman. The weight under claim constitutes the difference between the weight indicated on the Warrant and the weight on the signed Weight Note by the warehouseman at the time of release from the warehouse.

Coffee is weighed just before release after sale. The warehouseman is responsible for the weight shortage.

Weight shortages could result from spillage, tare weight inconsistency, sampling after the warrant has been raised or long-term storage due to low humidity and subsequent moisture loss. All documents for the purpose of any claim must be printed.Coffee stored longer than 8 weeks before sale must be reweighed.

CHAPTER 15

EXPORT PROCESS

EXPORT QUALITY PROCESS

Once buyers have bought their coffee, preparation for export begins for bulk coffee (blends), single origin and micro lots.

Buying sample

The first control point is to compare the Pre-offer sample with the Buying Sample. This is to verify that the lot purchased at the auction is representative of the Buying Sample.

Trucked (Railed sample)

The next control point is to compare the sample of the actual coffee received in the buyer's warehouse (Arrival Sample) with the Buying sample. This is to ensure that the coffee received in the warehouse for export is what was purchased, and no mistake has been made between the marketing agents' warehouse and buyers' warehouse.

Once the buyer has verified that he has the correct sample in his warehouse, coffee is allocated to the various contracts and preparation for shipment begins.

For single origins and micro lots, the buyer just needs to complete the markings according to the customers' requirements as most of the time these coffees do not need further preparation.

Bulk Shipments.

When it comes to the bulk shipments, the buyer will allocate coffee for a shipment and prepare what is called in the industry a Hand Bulk (a small representative laboratory bulk sample). This sample is taken from the various lots of coffee that constitute the Bulk. This sample is evaluated before the Bulk lot is assembled.

Bulk loading of Coffee into a 21,600 kilograms bulker bag through blowing. This is one of the packaging methods of coffees for shipment. At destination the coffee is directly blown into a silo. It is the most efficient and economical way of packaging and shipping coffee for big coffee roasters with silo storage.

Once the Hand Bulk sample attains the required quality in cup and green bean visual appearance, the actual commercial bulking of the consignment is carried out. The Bulk is either done manually or by machine, on a pro rata basis depending on the quantity of the bulk, the production capacity of the bulking machine or available space on the bulking floor. The ideal floor for manual bulking is a wooden floor with the bulking (mixing) done using wooden shovels. Metallic shovels could impart an off flavour to the coffee.

The last quality control point is comparing the Bulked Shipment Sample with the Hand Bulk Sample. The two must be the same in cup, appearance, and screen size. The Bulked Shipment Sample often tastes better than the Hand Bulk Sample.

Once the quality is verified the consignment is ready for shipment. Bulking coffee ensures that the quality of a shipment is homogenous. Roasters expect a homogenous coffee of consistent quality for their coffee blends.

Blending at origin enables coffee traders to put together a large quantity of standardized quality, good cupping coffee at a good price. This is achieved through mixing of different or similar grades (sizes) of coffee with similar cup quality and appearance. For example, a blend of AA and AB grades or purely an AB grade can be blended depending on requirements.

The number of cups tasted per bulk is determined by the quality of the coffee and the number of bags in a bulk. Lower quality coffee with many bags in a bulk will require the tasting of more cups e.g., Mbuni (unwashed coffee) and grinder quality coffees. Two cups are enough for main grade coffees. The exporter can decide the number of cups to be tasted per Bulk.

It worth noting at this point that all stock lot samples sent to customers for approval are samples made from the hand bulk samples of the intended shipment and should bear a reference number.

Kenyan Traditional Bulk Blends.

The quality of the Bulk quoted starts with the grade/s (screen size) and then the quality classification (mainly the cup and the green bean appearance).

FAQ Quality

A main Kenyan bulk coffee blend is FAQ which stands for "fair average quality". These coffees range from class 4 and class 5 in the classification category. Coffees in this category are quoted as AB or AA FAQ on the International market. FAQ quality coffee will have some minor defects, medium acidity, medium body, and flavor.

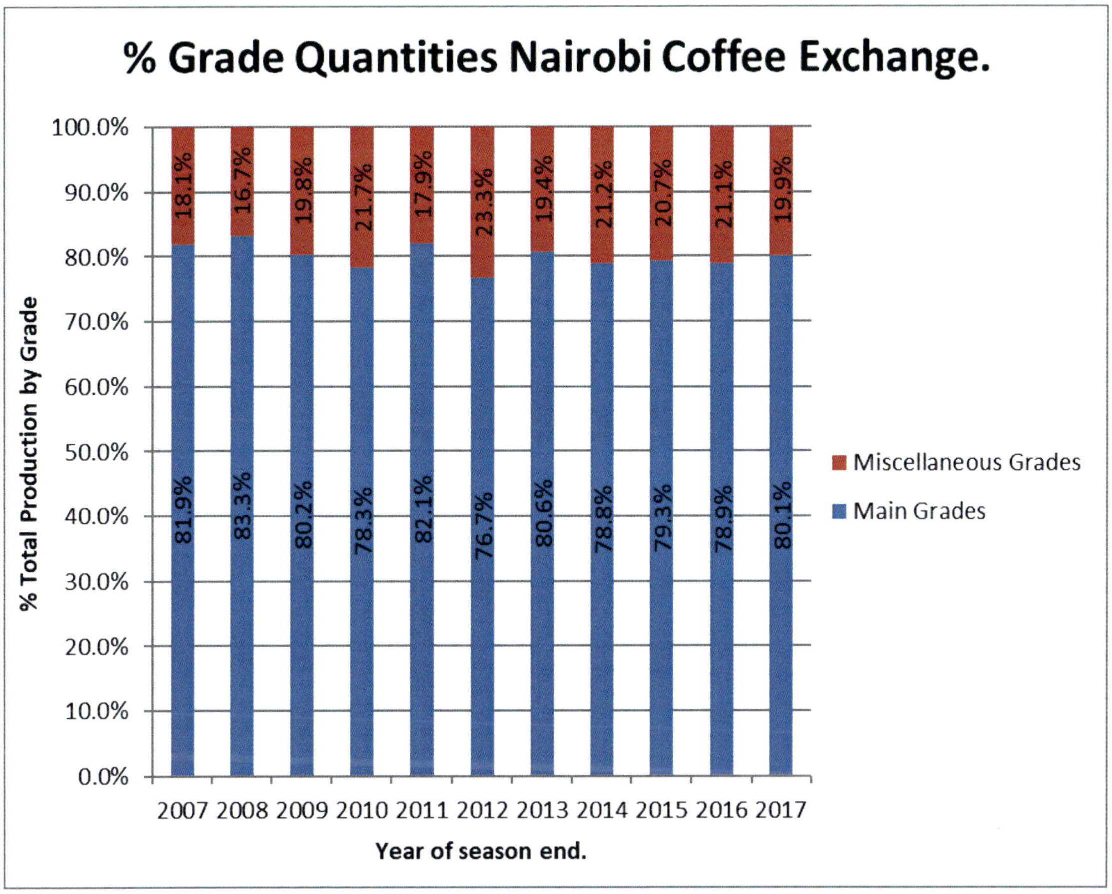

The above graph represents the general quality of coffee produced each year between 2007/2017. Main grades range from Class 1-6 while miscellaneous range from classes 6-10. (Class 1 is the best while class 10 is the poorest)

Plus Quality

Plus quality is Coffee quoted as AA or AB Plus and has bright center cuts, is uniform and solid, blue green or grayish green in color with barely any defects. The cup has a bright acidity, good body, and complex flavor with notable Kenya sensory notes of floral, citrus, raw sugar, and honey.

Top Quality

Top quality coffee is quoted as AA or AB top and has no defects, is bluish or grayish green in color, is a solid bean, and has white center cuts. The cup possesses bright acidity, good body, and a delicate well-balanced cup of fine complex flavor with floral, citric, black currant or honey notes. Top quality coffee is mainly found in grades AA, AB, PB and in limited quantities in C grade.

These lots can be offered as micro lots, single origin or as blends. The higher the quality the less available are the quantities. Plus, and the top quality would fall within the specialty category.

Other Blends

Some traders and their customers develop their own blends in additional to the "Traditional Kenya Blends" An example is AB minus.

Loose coffee bags being stuffed into a container. Paper liners protect the coffee from effects of condensation. A normal loose load is 320 bags with a maximum capacity of 330, 60-kilogram bags

Micro Lots

Micro lots are basically small lots of coffees grown under the same micro climatic conditions, picked, and processed in same way as one lot. They are mainly in the "Plus" and "Top" quality categories. Every bag of a given micro lot has the same quality characteristics.

Single Origin

The term single origin refers to a coffee from one country. It is mainly used in reference to "Plus" and "Top" qualities which fall under speciality coffees and are sold as individual grades.

These coffees are sold without blending and the key aspect is that the coffee is traceable back to the factory.

Types of samples for internationally traded Coffee.

1) **Pre-shipment sample**: This is a sample sent to customers for approval against a contract before the actual shipment and is a representative of the actual coffee and quality of the stock lot earmarked for shipment.

2) **Type sample***:* This sample is also known as a Pilot (Reference) sample which represents a given quality of coffee established between an exporter and their customer. It is not a stock lot but represents the quality of coffee to be offered for shipment.

3) **Offer sample**: This is a sample representing stock held by an exporter/importer which is offered to a customer without their request.

4) **Shipment sample***:* This sample represents the actual coffee shipped and is drawn just before shipment. It may be drawn by the shipper or a third party.

*5)***Landing/arrival sample** This is a sample representing the actual coffee shipment on arrival and is drawn by the importer or the roaster.

Green coffee being bulked manually

CHAPTER 16

TRANSPORTATION AND SHIPPING

A vessel loaded with containers ready to sail

Kenya as a country has two major modes of transport, road and rail. Road transport has been a dominant mode of transport in the industry as it is faster, reliable, and convenient.

During the years when all coffee was marketed by the Kenya Coffee Board, all the coffee in Kenya was transported only by railway to the port of Mombasa or authorized warehouses.

Rail although being the safest mode of transport, became inefficient and unreliable. This led coffee buyers opting for road transport. Most of the coffee growing areas now have feeder roads linking them to the main roads. Coffee buyers truck their coffee to their warehouses and to the port located at the coastal town of Mombasa, 488 km from the capital city of Nairobi.

The recently constructed standard gauge railway is once again becoming a major transport system. Its success will depend on cost effectiveness and efficiency. Rail offers greater security as once coffee is handed over at the inland port in Nairobi it is placed under the custody of the Kenya Port Authority and Kenya Railways, absolving the exporter from any responsibilities and risks along the way.

Warehousing: The Kenya Coffee Directorate is entrusted with responsibility of ensuring warehouses used for storing coffee meet the required standard. A coffee warehouse roof should be high, insulated on the inside and ventilated to prevent excessive heat in the warehouse. It should be clean with coffee bags placed on pallets to allow air flow at the bottom of the stack and prevent bags at the bottom from water damage in the case of leakage.

The warehouse should be secured, and coffee stacked in a way that ensures safety during operations including inspection, with pathways along the stacks to allow easy movement of personnel.

CHAPTER 17

TRACEABILITY OF KENYAN COFFEE

Traceability is a new requirement in the global market and is gaining popularity in the coffee speciality market.

Traceability should generally guarantee fair prices to producers at origin, an ethical standard of production and processing, environmental sustainability and for consumers safety and authenticity of product.

The Buyers of traceable products are willing to pay more to compensate for these advantages. Traceability provides for differentiation of lots of coffee or grades produced in different regions and farms processed within a given production year. It is also a way of validating claims or coffee information provided by the exporters.

The Kenyan coffee marketing channel enables traceability of all coffee lots sold, whether as micro lots, single origins, or bulk consignments. Traceability starts at the wet mill (factory level) for cooperative farmers and farm level for plantation (Estate) farmers.

Cooperatives mix individual member's cherry at the factory, which is processed, milled, and sold as one bulk. Coffee from medium and large estates (plantations) is traceable back to the grower. using the Farm Name, Reference, and Outturn Number in the Outturn Statement.

The Outturn Statement includes delivery week, abbreviation of the miller code, the consignment serial number, factory name (in case of Cooperatives), coffee district code, Society Code and Factory.

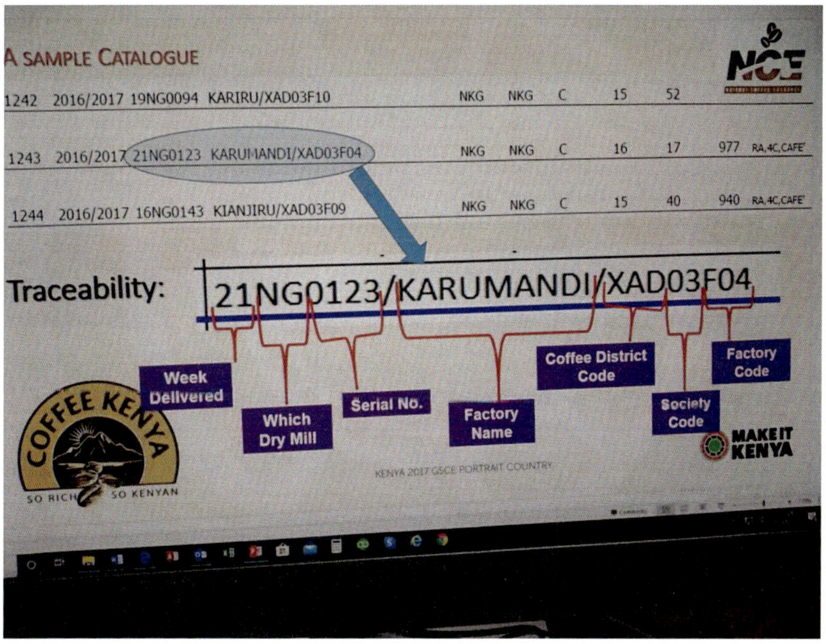

Courtesy of Nairobi Coffee Exchange

Kenya coffee traceability documents along the marketing chain

Delivery Slips

Delivery slips are issued upon delivery of parchment coffee to the mill. Delivery slip indicates total number of bags delivered and is issued after every delivery. The Delivery slip is identified by its Outturn Number.

Parchment Bulk Statement

This is issued when coffee is bulked together before milling. After Bulking, the Bulk acquires another number with details of all the parchment coffees that are in the Bulk for traceability and payment purposes.

Small Bulk Statement

It is a bulk of different small lots of green coffee of the same quality. The bulking is done to facilitate marketing of coffee mainly at the auction. The bulk is made up of clean (green) coffee with lots of a few kilograms. It would not make sense to list these lots in the Catalogue for sale at the Auction.

The report details different outturns bulked together, their grade and their corresponding weight. This facilitates both traceability and payments to different farmers with coffee in the Bulk. Small Bulks mainly constitute grades like E (Elephant) PB (Pea berry) and T grades which are normally in small percentages.

Payment Statement

The Payment Statement indicates the Outturns of the coffee being paid for, the Sale Number in which the coffee was sold at auction, the number of bags and their corresponding weight, the grades of the coffee, the gross value of the coffee, the standard deductions, and the net payment to the farmer.

These coffee bags in storage tagged with markings for traceability have multiple certifications

CHAPTER 18

ECONOMIC SUSTAINABILITY OF KENYAN COFFEE

There are various certification organisations in Kenya, namely: UTZ-Rainforest alliance, Fair Trade, 4C (common code for coffee communities) and Café Practices. The merger of Utz and Rainforest Alliance will save money for farmers that previously required both certifications. Certifying organisations mainly focus on social, economic, and environmental practices at farm level.

Coffees sold under certifications attract a premium price above coffee of similar quality. Certification is meant to ensure coffee sustainability providing better earnings for farmers. However, many small farmers cannot afford the certification fees while many farmers that can afford the fees, are unsure which certification will guarantee the maximum return. Having multiple certifications is costly.

Despite the good intentions of Certifying Organisations, Kenya coffee production has been declining. The private sector through leading trading houses and affiliate organisations have been involved in sustainability programmes without any real impact on overall Kenyan coffee production. There is a need for government effort, to keep reforming the sector in response to emerging challenges and marketing dynamics.

Coffee growing in Kenya was pioneered by the Scottish mission. Kenya produced 1000 tonnes of coffee in 1923. Production increased to 11,000 tonnes in 1928. This was shortly after the formation of the Coffee Planter's Union in 1926. In 1933 Kenya produced 12,848 tonnes of coffee, while in 1940 it exported 13,000 tonnes. Between 1940 to 1984 Kenya coffee production had grown to 128,926 tonnes. Production grew at an average annual rate of 6.6 % between 1963 and 1988. Currently the area under coffee is estimated at 115,570 hectares down from 170,000 hectares.

Kenya Coffee production Between 2012/2013 to 2017/2018 in metric tonnes

Year	**2012/13**	**2013/14**	**2014/15**	**2015/16**	**2016/17**	**2017/18**
Estates	13,773	16,805	14,808	15,267	14,151	10,997
Cooperatives	26,092	32,652	27,230	30,758	24,468	30,381
Total	39,865	49,475	42,038	46,121	38,620	41,375

The table above shows there has not been significant changes in production since 2012/13

Source ICO coffee Profile (Kenya)

Factors contributing to the decline in Kenyan Coffee Production

Price instability, and low earnings have led to farmers uprooting or abandoning coffee in favour of more profitable crops. This is a global phenomenal affecting coffee farming all over the world.

Unlike countries like Ethiopia and Brazil who consume about fifty percent of their production, Kenyan domestic consumption is estimated at a mere 5 %. This is not enough to support prices internally when world market prices dip.

Conversion of coffee growing land to real estate is prevalent especially in and around Nairobi County where population growth has heightened demand for housing which offers far higher returns to farmers than coffee growing.

Coffee marketing dynamics, limited research, inefficiencies by management, have all limited the realisation of the full potential of Farmers' Cooperatives. The collapse of the Kenya Planter's Cooperative union is an example.

The younger generation is also not interested in coffee farming due to the hard work that goes with it. Subdivision of family farms into smaller unviable economic units and weather changes have also affected coffee production, especially with recent frequent droughts and unsynchronised rain patterns

Cash flow and price uncertainty associated with marketing dynamics is one of the main causes of decline in production as most farmers cannot afford to buy inputs when they need them. The "Cherry Advances Revolving Fund" proposed by the government is meant to address this problem.

According to the Kenya Agricultural and Livestock Research Organization, the highest cost of cherry production is approximately USD 0.51 per kilogram while the lowest cost is USD 0.097 per kilogram depending on variety and farming practices.

The collapse of the International Coffee Agreement in 1989 following disagreements by member countries on new supply quotas, destabilised coffee prices and subsequent local farmer's earnings. This raised the question as to whether the Kenya coffee industry was thriving due to protection from the quota system or due to competitiveness in production and management of the industry in general. Another factor affecting coffee production is the limited research supporting the industry.

Previously scraped levies on coffee farmers are being reconsidered. These levies are intended to fund research which is critical to the industry. Leaf rust and coffee berry disease remain a major challenge affecting both yield and cost of production. Farmers must spray their coffee to control these diseases. Research to find solutions to combat these diseases would translate to better earnings. A good example is the introduction of Batian which is a new variety producing higher yields and is resistant to coffee leaf rust and coffee berry diseases.

Coffee Cooperative Societies have a role to play in the industry. Individual farmers are organised into Cooperatives which allow farmers to pool their resources to provide processing and management services for members further up the value chain. Prudent management of cooperatives is critical in sustaining coffee farming. Stand-alone farmers with uneconomic coffee enterprises should be encouraged to join cooperatives. Cooperative produce more coffee compared to large scale farmers.

Possible ways to improve earnings for farmers

For economic sustainability Kenya needs to produce coffee with high yield per hectare at a lower cost to achieve better margins. Reducing farmer's costs for energy by exploring solar energy, bulk purchase of fertiliser and insecticides, Subsidy or zero tax on imports for farming inputs are examples.

Planting a high yielding variety like Batian which is resistant to coffee berry disease and leaf rust will help lower cost of inputs by eliminating the cost of spraying while at the same time improving yield per hectare. The Government should facilitate bulk procurements of inputs by societies or unions for their members at competitive prices.

There is need to explore other coffee processing methods for speciality market that guarantee better earning. This will require resources and expertise.

According to world coffee research, most African countries produce less than 500 kilograms of green coffee per hectare compared to other countries that are producing between 1000 kilograms and 2500 kilograms of coffee per hectare. Kenyan farmers need to increase production per hectare. The Kenya national average coffee yield is estimated at 302 kilograms per hectare for Cooperatives Societies and 556 kilograms per hectare for Estates (large scale farms)'. Small holdings work against achieving economies of scale. Formation of Co-operative Societies partly solve this problem.

Average Production Yield for African Countries between 2010/11 to 2013/14 Source ICO (114-) (ICO Kenya coffee profile).

Growing Country	Estimated area under Coffee In Hectares	Estimated yield green beans (Clean coffee) per Hectare in kilograms
Kenya	110,000	412
Tanzania	229,000	216
Uganda	282,284	708
Rwanda	42,000	385
Ethiopia	509,000	800

Maintaining the Kenya Coffee Auction system offers the best price discovery mechanism through competition in an open and transparent manner guaranteeing improved earnings to farmers and encouraging production of high-quality coffee.

The Coffee Directorate needs to collect and maintain credible data on coffee crop husbandry, production, and marketing statistics to be able to manage the industry in a timely manner and intervene where necessary. It could also explore other potential coffee growing areas to increase production.

Real growth in local coffee consumption can also help support coffee prices. Currently Kenya consumes only around 5 % of its total production. An effort by the government in the early 90s to support local consumption growth through subsidies was not successful. Coffee meant for local consumption found its way to the export market.

CHAPTER 19

VALUE ADDITION

Value addition leads to better earnings for coffee farmers. Selling Kenya coffee in green bean form rather than parchment has ensured better earnings for Kenyan farmers. It is easier to assess the quality of the coffee in green bean form realising more accurate grading and assessment of quality and subsequent improved prices.

Farmers in origins that sell their coffee in parchment or cherry form, especially Arabica, at the farm gate do receive prompt payment, but the price may not reflect the true value of their coffee. The Farm Gate Price is based on general parchment or cherry quality ignoring any inherent quality disparity between different batches.

The government has been exploring exporting roasted coffee in the international market. This is a different terrain altogether which will require understanding the target market complexities. Different markets require coffee roasted to different levels reflecting how the coffee is consumed. Some markets are mainly milk-based coffee drinkers while other markets have their coffees as short and long black. Most milk-based coffees are blends of different origins. Another challenge of exporting roasted coffee from origin is the freshness of the coffee when it hits the market, and its shelf life.

KENYAN COFFEE PRODUCTION STATISTICS
KENYA COFFEE PRODUCTION DATA

Annual Quantity (60 Kg Bags) and Highest Price Paid in $US for Kenyan Grade AA Beans

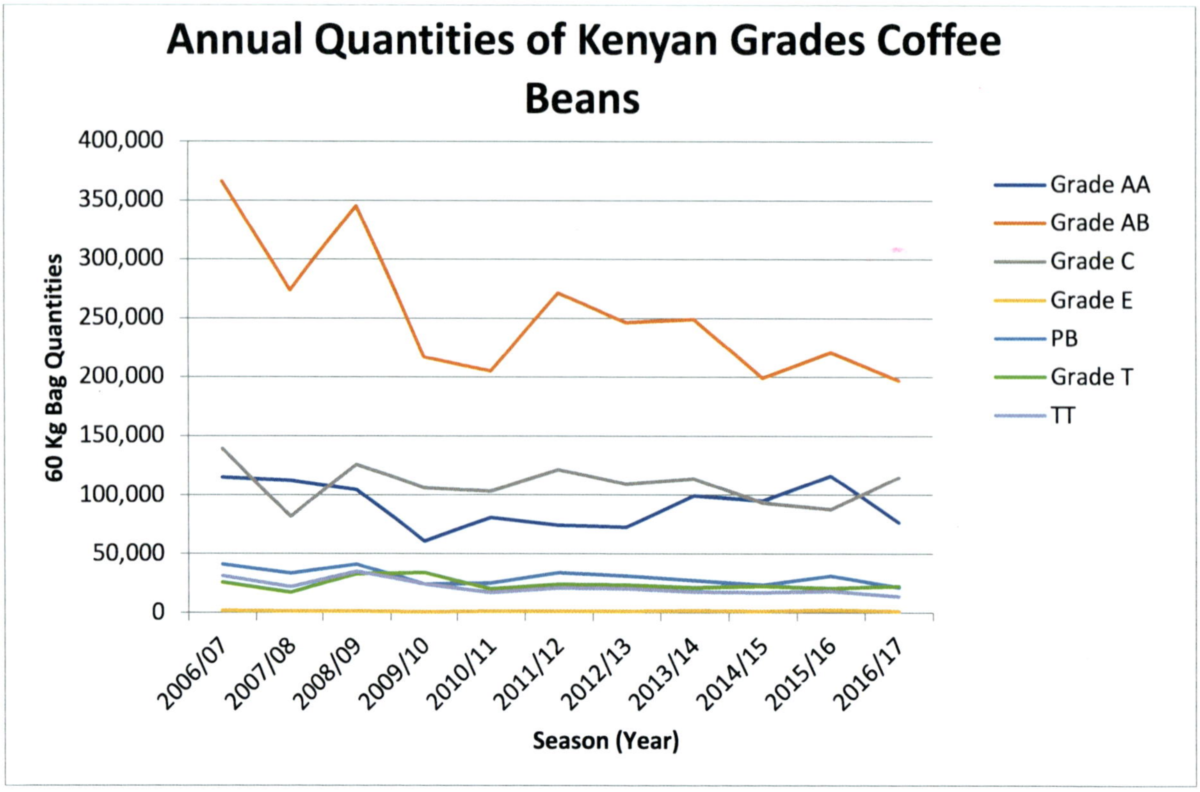

Courtesy of Nairobi Coffee Exchange

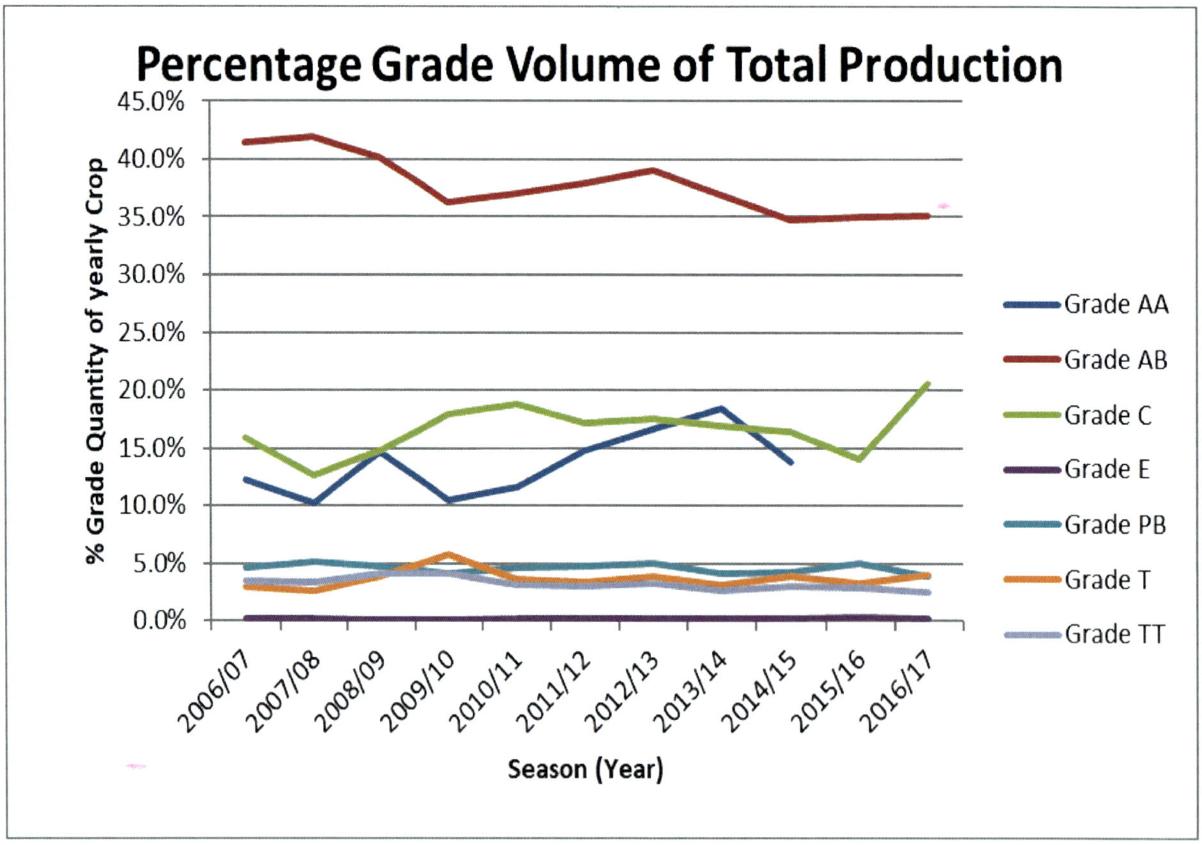

Courtesy of Nairobi Coffee Exchange

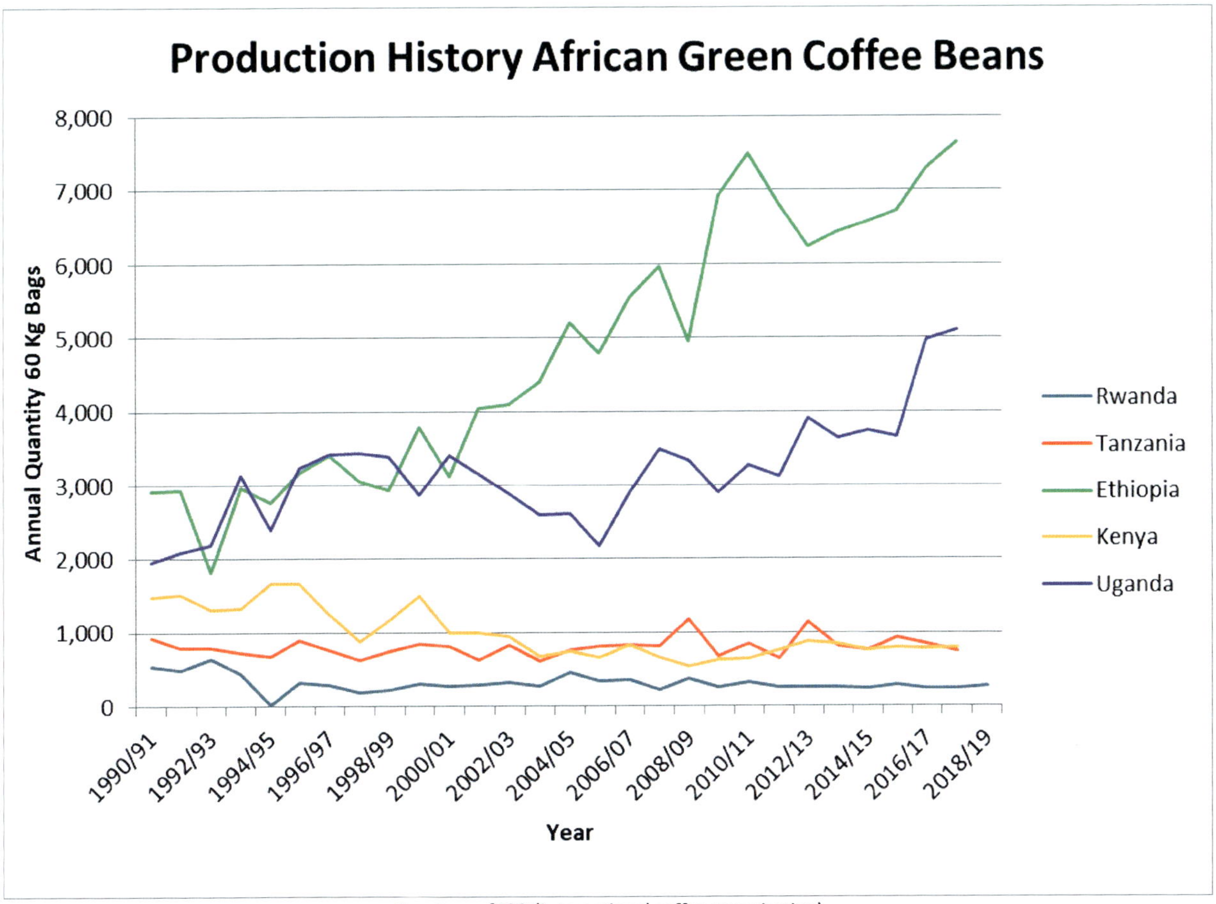

Courtesy of ICO (International coffee organisation)

CHAPTER 20

OTHER AFRICAN COUNTRIES

UGANDA COFFEE MARKETING SYSTEM

Uganda coffee is sold through direct sales with buying of either cherries or parchment at the farm gate. There is a proposal to introduce a voluntary Auction.

Uganda grows both Arabica and Robusta species. Robusta accounts for about 80% to 85% of the total production.

Arabica varieties grown in Uganda are SL 14, SL 28 which are Bourbon related and Ruiru 11. Uganda main Robusta varieties are Nganda and Erecta (ICC-124-8 2019). The estimated area under coffee is about 282,284 hectares. The estimated yield for Arabica is 500 kilograms per hectare and 600 kilograms per hectare for Robusta (Uganda Coffee development Authority fact sheet).

Coffee marketing participants include Small-Scale Traders, Farmers' Co-operatives, Processors and exporters.

Value addition in Uganda is dominated by the private sector. Smallholder farmers harvest and sun-dry most of their Robusta cherries (known as Kiboko) and sell it at the farm gate to small traders/ cooperatives or collectors for cash. Collectors in turn sell the coffee after hulling to big independent traders or to exporters and their agents.

Hulled ungraded green beans are referred to as FAQ (fair average quality) both for Arabica and Robusta.

Secondary processing also referred to as export grading, involves grading the hulled coffee to standard grades that meet international standards and cleaning and drying the coffee if not properly dried in preparation for export.

Some farmers pulp their Arabica coffee using hand pulpers and sell it as parchment. Bugishu farmers also sell cherry to private wet mills owned or managed by exporters. Exporters who own dry mills buy coffee directly from middlemen or farmers and process and grade, prior to export. Some exporters even own coffee farms.

One of the challenges facing the Uganda Coffee Industry is the marketing structure which encourages severe competition at different levels of the value chain pitting quantity against quality.

Dry processed (unwashed Arabica) is marketed as Drugar type (Dry Uganda Arabica). Washed Uganda Robusta marketed as Wur, Wet processed (washed Arabica) is marketed as Wugar type (Wet Uganda Arabica) and Bugishu (both semi and wet processed).

Bugishu Arabica is grown on the border of Uganda and Kenya around Mount Elgon by the Bugishu people and is in high demand. Uganda Smallholder farmers intercrop their coffee with food crops utilizing shade from crops like bananas.

Uganda coffee is sold on a FOT (free on transport) basis Kampala, FOB (Mombasa Port), CIF (Cost, Insurance and Freight).

The Uganda Coffee Classification is based on bean size, region (e.g., Bugishu) and number of defects.

Uganda Robusta Main Commercial Grades, description, and defects

Grade	Screen size	Defects
Screen 18	92 % retained by screen 18 and 0 % through screen 15	Maximum 7 % in weight of defective beans
Screen 15	90 % retained by screen 15 and 0 % through screen 12	Maximum 12 % in weight of defective beans
Screen 12	85 % retained by screen 12	Maximum 20 % in weight of defective beans

Robusta should have a maximum 12.5 % moisture content, clean cup and not more than 0.1 % of foreign matter.

Uganda Bugishu and Other Arabica Export grades

(Source: Arabica Coffee handbook -UCDA 2019)

Grades	Screen Size	Retention %	Total Defects (By count)	Remarks
Bugishu AA	17	90	5 %	
	16	8		
	15	2		
Bugishu A	16	90	7 %	3 % retained on screen 17
	15	8		
	12	2		
Bugishu A +	17	60	5 %	Mixture of AA &A from Bugishu and Kapchorwa areas
	16	30		
	15	8		
	12	2		
Bugishu B	15	90	10 %	
	14	8		
	12	2		
Bugishu PB	15	80	10 %	Tolerance of A=0-2 %
	14	15		
	12	5		
Bugishu AB	17	8	7 %	A mixture A & B
	16	60		
	15	30		
	14	2		

Grades	Screen Size	Retention %	Total Defects (By count)	Remarks
Bigishu CPB	15	60		Should have zero A and AA
	14	38		
	12	2		

Bugishu coffee must have good appearance relative to its grade, solid. beans with a clean cup and 0.1 % of foreign matter.

New grades introduced in the washed Arabica category are Mt Elgon AA, Mt Elgon A+ and Mt Elgon A. Although these three grades overlap other grade categories from the Bugisu region in screen size and description, they are specifically from the Bugisu and Sebei areas around Mt Elgon. Their quality criteria specify no primary defects, five to eight secondary defects, zero to three quakers in 100 grams of roasted sample with a minimum of 80 points on the SCA scale.

Coffee Marketing Value Chain (ICO Uganda country coffee profile ICC-124-8)

Uganda coffee marketing channel offers various alternatives. Some farmers sell to middlemen, while others sell directly to coffee buying storemen. Famer's organisations (cooperative societies) sell directly to exporters while others export the coffee themselves.

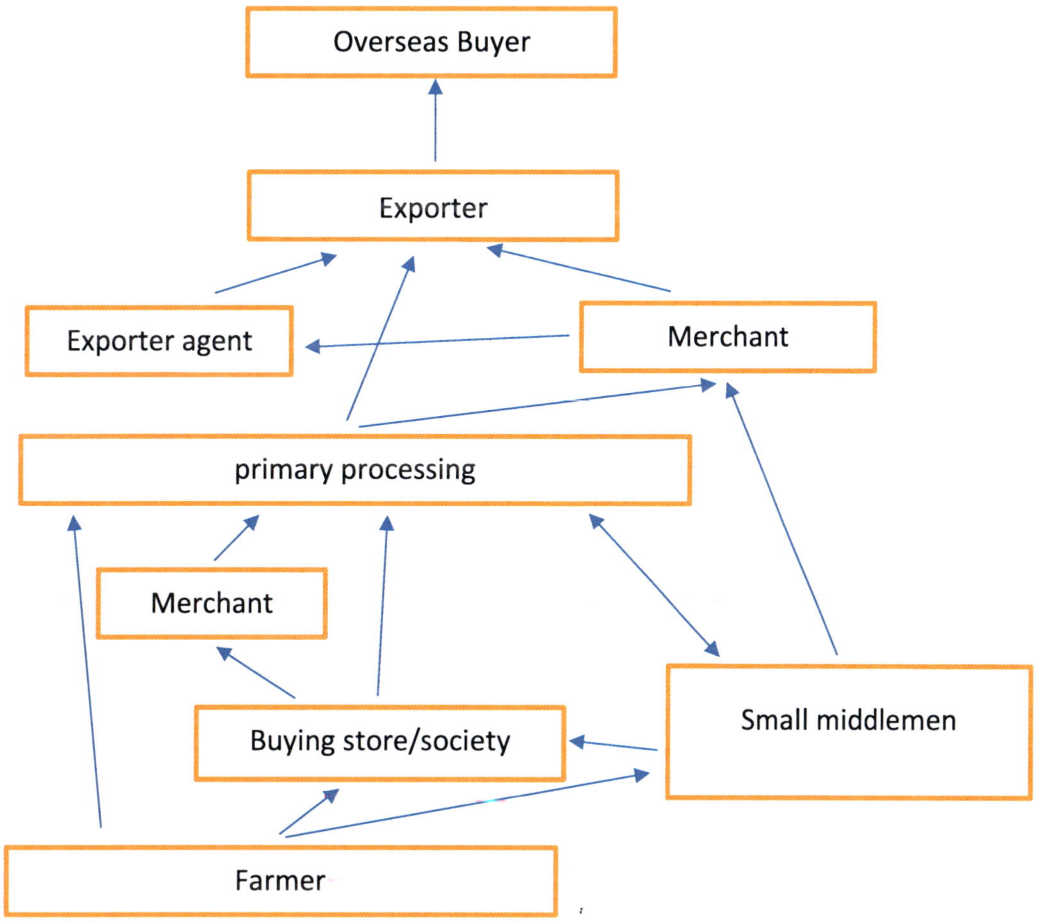

Chapter 21

TANZANIA COFFEE MARKETING SYSTEM

Tanzania grows Bourbon and Kent Arabica varieties. Arabica coffee is grown on the slopes of Mount Kilimanjaro and Mount Meru in the Northern areas under the shade of banana trees. Arabica coffee is also grown in the southern highlands of Mbeya and Ruvuma regions.

Arabica is grown at altitudes of between 1000 meters to 2500 meters above sea level while Robusta is grown along Lake Victoria in the Kagera region at an altitude of between 800 meters and 900 meters above sea level.

Arabica coffee grown around Mount Kilimanjaro has a similarity in cup characteristics to Kenyan coffee but attracts lower price differentials. About 98 % of Tanzanian Arabica coffee is wet processed. Tanzanian coffee grading is based on size, density and defect counts. Arabica constitutes 70 % of the total coffee production while Robusta is about 30 %.

Harvesting starts in July and ends in December in both the North, and Southern Tanzania, while in Western Tanzania harvesting starts in May and ends October. It is estimated the area under coffee is 265,000 hectares.

The Tanzanian coffee marketing system has evolved from a centralized auction, where all coffee traders purchased coffee for export. Further changes allowed traders to buy parchment from the farm gate alongside the Auction system. The buyers would mill their coffee (already bought as parchment) and offer it at the auction before export. This meant they (buyers) would bid on both their own coffee and other coffees at the Auction for export. It was a government policy for all coffee to go through the auction system prior to export.

More changes established two marketing channels with the option of selling coffees scoring class four and above through the auction system and selling coffee between class one and three directly to overseas buyers.

In 2019 Auctions were decentralised from Moshi to the coffee growing regions with provision for online biding. Currently Tanzania allows direct sales of coffee as an alternative to auction. Private buyers with processing facilities can buy parchment and dry cherry directly from the farmers.

In further reforms the Tanzania Coffee Board reduced the number of days for payment settlement for coffee purchased at the auction from seven to four days after purchase. This fast-tracked payments to the farmers. The Coffee Board also introduced a special catalogue for certified coffee to include Organic, Fairtrade, Rainforest Alliance/Utz and Café Practise and all speciality coffees with scores of 80 points and above on the Q grading scale. The expectation is that potential high prices scored by these coffees will motivate farmers to improve their quality and earnings.

Tanzania Arabica Commercial grades

Grade	Screen Size
AA	Minimum 90 % retained by screen 18 and maximum 98 % to 100 % retained by screen 17
A	Minimum 90 % retained by screen 15/16 and maximum 2 % retained by screen 14
B	Minimum 90 % retained by screen 15/16
C	Maximum 10 % retained by screen 14
PB	Minimum 95 % retained by pea berry screen 14
AF	Minimum 90 % retained by screen 17 and 98 % to 100 % retained by screen 15/16
E	Minimum 90 % retained by screen 18 and 100 % retained by screen 15
F	Constitutes broken beans and floaters from other grades.

Source: Aziende Ruinite Caffe

Tanzania Robusta Commercial grades

Grades	Screen Size	Defects
Grade 1 Robusta Superior	95 % above screen 16	15 defects,1% partial blacks, and insect damaged beans
Grade 2 Robusta FAQ	95 % above screen 14	40 defects ,1.5 % blacks
Grade 3 Robusta special (SUG)	95 % above screen 14	60 defects,2.5 % blacks, husks, and pods
Grade 4 Robusta UG	90 % above screen 13	160 defects, 60 being blacks, foreign matter 1-2 % including pods
Grade 5 Triage	Nil	Contains 1% husks

The Cup must be clean without any musty off flavour. Tanzania exports Instant Spray Dried Coffee as a value-added product. The majority shareholding of this factory is in the hands of farmers' cooperative unions.

Tanzania Coffee Marketing Value Chain

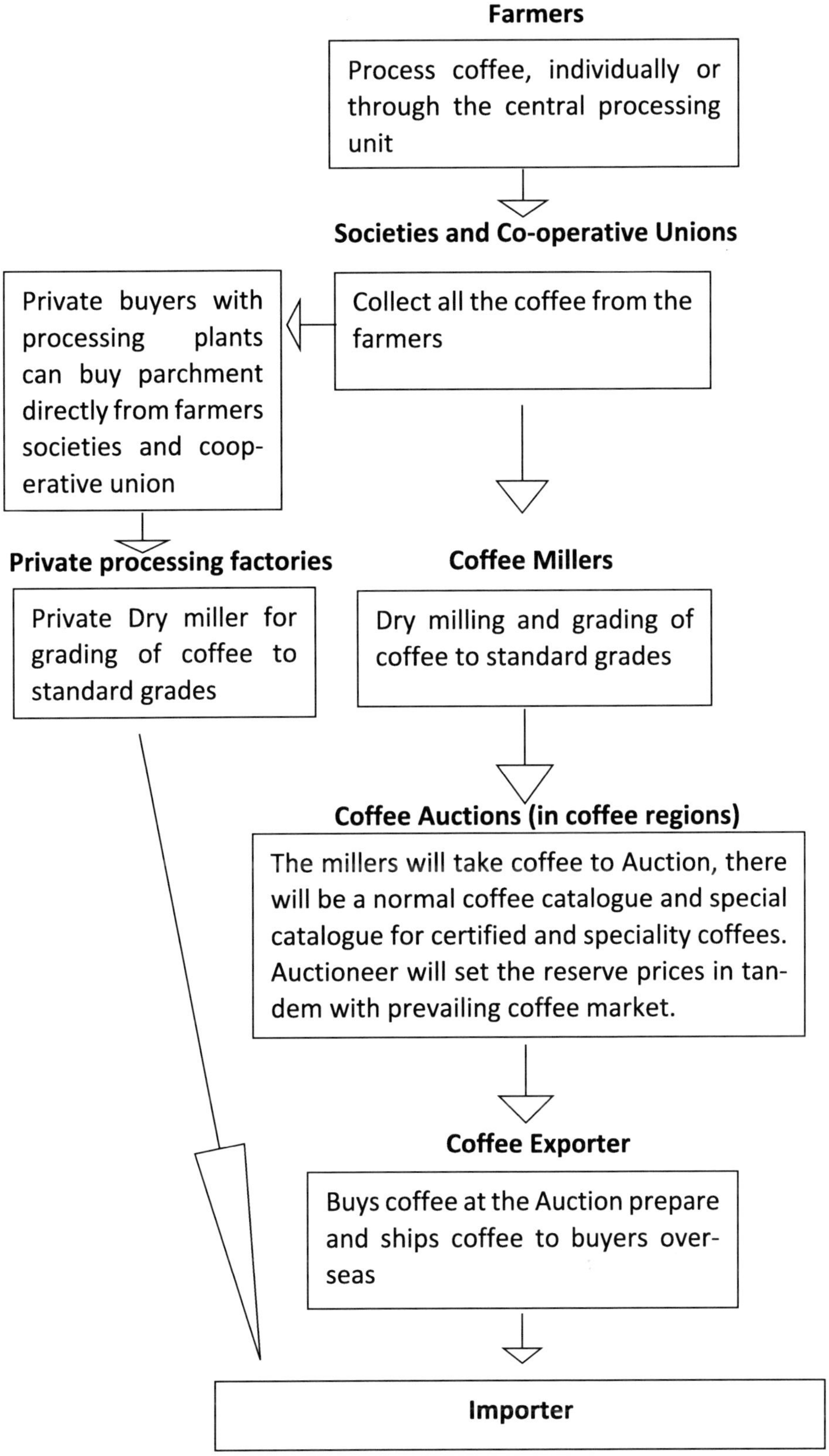

CHAPTER 22

RWANDAN COFFEE MARKETING SYSTEM

Rwanda mainly grows the bourbon Arabica variety, Caturra and Catuai. With one coffee production season in a year, harvesting starts in February in the Western Province on the shores of Lake Kivu and in March in other parts of the country, ending in August. Rwanda produces both wet and dry processed coffees. Prior to changes in the coffee industry, export of coffee was carried out by two companies. However, changes in the industry ushered in an active role by both the private sector and coffee growing associations.

Coffee is sold in both parchment and cherry form, with prices set every week. Rwanda coffee is grown at an altitude of about 1900 meters above sea level. The estimated area under coffee is approximately 42,000 hectares. Rwanda coffee is graded according to size, defects, odour, colour, and cup.

Green Coffee is graded (classified) into the following grades.

Grade	Score	Defects	Screen Size	Cup
Super speciality grade	90 -100 % Points	No primary defects, no more than 5 full defects, No Quakers are permitted, With moisture content of 10-12.5 %	No more than 5 % below specified screen size	Must have at least one distinctive attribute in the body, flavour, aroma, or acidity. The cup must be free from cup faults and taints.
Speciality grade	80- 90 % Points	No primary defects permitted with no more than 5 full defects. No Quakers are permitted. The moisture content should be between 10-12 .5 %	Maximum of 5 % below specified screen size	Must have at least one distinctive attribute in the body, flavour, aroma, or acidity. The cup must be free from cup faults and taints.
Grade 1	70-79 % Points	No more than 9-23 full defects, No Primary defects allowed.	Maximum of 5 % below specified screen size	Must have at least one distinctive attribute in the body, flavour, aroma, or acidity. The cup must be free from cup faults and taints.

Grade	Score	Defects	Screen Size	Cup
Grade 2	60-69 % Points	No more than 24-86 full defects	unspecified	unspecified
Grade 3	50- 59 % Points	More than 86 full defects	unspecified	unspecified

Semi Washed (Sample size is 350 gram)

Grade	Score	Defects	Screen Size	Cup
Grade 1	70-80 % points	No more than 23 full defects, Primary defects are permitted. only 3 Quakers permitted. The moisture content should be between 9-12.5%	Maximum of 5 % below the specified screen size	Must have at least a distinctive attribute in the body, flavour, aroma, or acidity. The cup must be free from faults and taints.
Grade 2	55-70 % points	No more than 30 full defects, Primary defects are permitted. Only 3 Quakers permitted. The moisture content should be between 9-12.5 %	Maximum of 5 % below the specified screen size	Must have at least a distinctive attribute in the body, flavour, aroma, or acidity. The cup must be free from faults and taints.
Grade 3	40-50 % points	No more than 50 full defects	unspecified	1 to 2 defective cups
Grade 4	Below 40 % points	More than 80 full defects	unspecified	More than 2 defectives cups

Source: Rwanda Coffee Grading system (National Quality standards ICC 122-12)

Robusta

Fully washed Robusta is graded as grade one while semi washed Robusta is graded as grade two.

Rwanda Coffee Value Chain

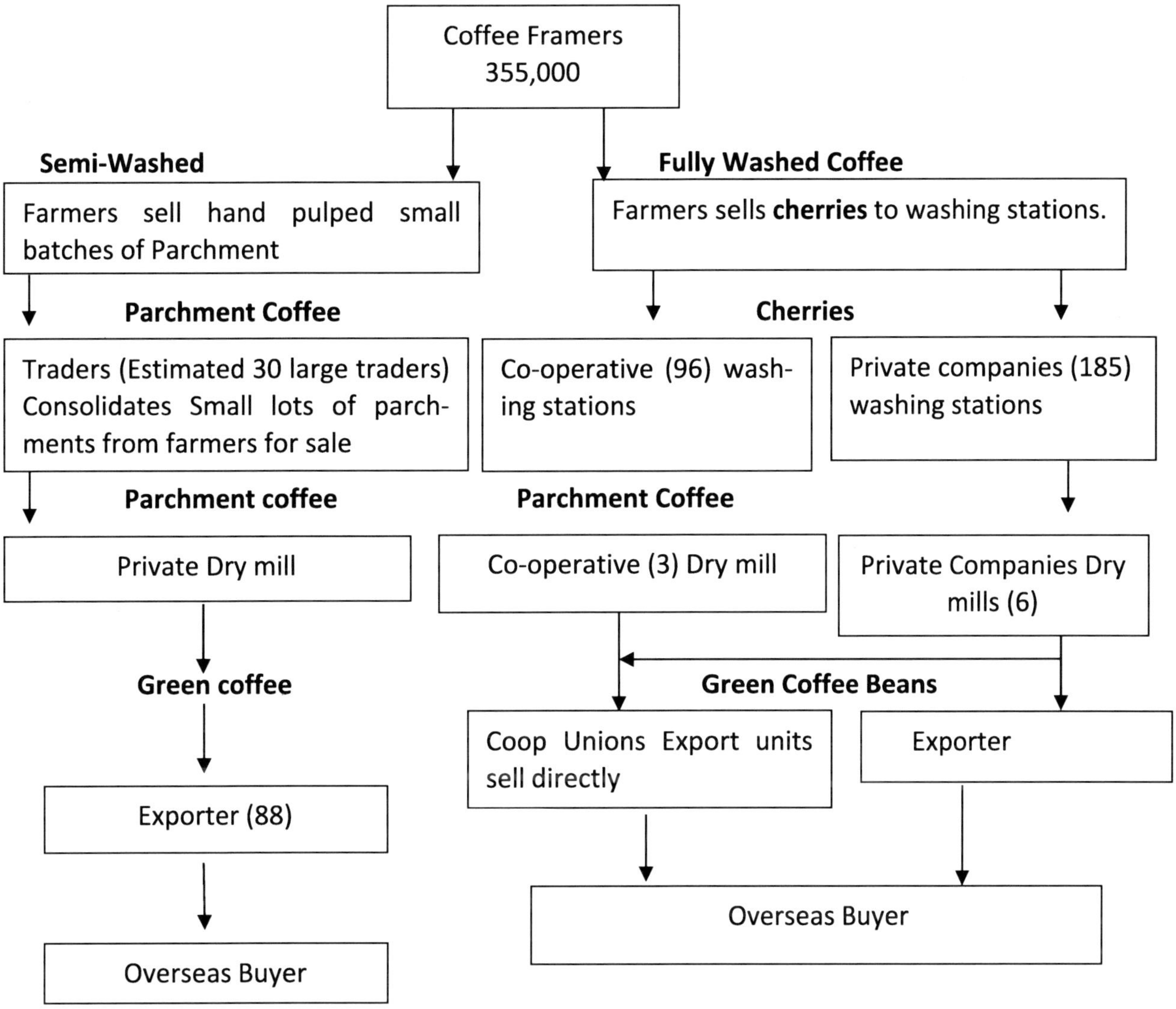

Source: Adapted from Agri Logic Value Chain (originally adapted from TWIN andTMEA) 2018

Cherries are delivered to buying centres by Collectors and ferried by truck to a washing station. Farmers near washing stations sell their coffee directly to the washing station.

Washing stations are owned by cooperative associations, secondary processors, and private companies. Some cooperative associations also have dry mills. There has been a lot of investment in washing stations by the government. Washing Stations have replaced hand pulping machines. Cherries that do not meet the quality threshold are hand pulped and sold to middlemen as parchment for use in semi washed coffees.

Quality is controlled both at the collection centre and the washing station. Collectors are paid by the washing stations who in turn pay the farmers.

CHAPTER 23

ETHIOPIAN COFFEE MARKETING SYSTEM

Ethiopia is the origin of Arabica coffee and the largest coffee producing country in Africa. Ethiopians are a coffee drinking nation, consuming almost half of their production.

Most of the Ethiopian coffee is grown at middle to high altitudes yielding a hard bean with an intense flavour. Berry, citrus, and floral fruity flavour notes set Ethiopian coffee apart. Each coffee growing region has its unique flavour notes adding to complexity of flavour in a bulk. Ethiopia grows Arabica species only. Both sun dried and wet processed coffees are found in large quantities.

Small holders produce about 95 % of Ethiopian coffee in environments ranging from forest, semi forest, garden, and plantations.

One of Ethiopia's main coffees growing regions is Harar in Eastern Ethiopia. This region only produces sun-dried natural coffee with distinct blue berry notes. Constant demand for Harar coffees from Saudi Arabia keeps prices from this region slightly higher than coffee from other regions.

Sidama also spelled as Sidamo is a coffee growing region in Southern Ethiopia. Sidama produces both washed and unwashed coffees with complexity and intense fruity notes.

Another major coffee growing region is Yirgacheffe which is a small micro region within the larger Sidama region. Yirgacheffe coffee is grouped into a separate category due to its flavour profile and is well known in the speciality market for its delicate floral citrusy notes. Sidama and Yirgacheffe coffees normally command a price premium.

The Limu coffee growing region is found in the Southwest of Ethiopia. Limu coffees are characterised by mild acidity. Traditionally coffees sold as Limu are normally washed while unwashed coffee is sold as Jimma. Jimma also spelt as Djimmah is the largest source of unwashed coffees, produced in the Southwestern region of Ethiopia.

Nekempti also known as Lekempti represents coffee from Western Ethiopia. Traditionally coffee from this region is Sundried Natural characterised by large beans.

Ethiopian coffee can be procured through three main channels namely Exporters, Cooperative Unions, or Private Coffee Estates (farms). Coffee export businesses have been traditionally restricted to locals only. Exporters generally purchase their coffee from the Ethiopian Commodity Exchange, while Private Estates and Cooperative Unions with an export license can sell directly to the international market.

Ethiopia Coffee growing regions and types of coffee produced

Region	Types of coffee produced
Haral Region	Only produces Sundried Natural
Sidama Region	1.Washed (Wet processed) 2.Unwashed (Natural)
Limu Region (Traditionally Coffees sold as Limu are normally Washed)	1.Washed coffees (from Limu region) sold as Limu 2.Unwashed coffees (from Limu region) sold as Jimma
Yigarcheffe (is Micro Region within Sidama)	1.Washed (Wet processed) 2.Unwashed (Natural)
Lekempti Region (Originally producing Naturals)	1.Producing Unwashed (Naturals) 2.Washed (Wet processed)

Primary cooperatives generally channel their coffees through Cooperative Unions. This helps most farmers who have small parcels of land to pool their coffees together for export under the cooperative umbrella. Cooperatives have their own washing or drying stations. Coffees produced from cooperatives and private farms are exempted from going through the Ethiopian Commodities Exchange (ECX) and are traceable down to the producing district and community level respectively. Traceability assures consumers that farmers benefit from their purchases.

Offering of coffee through the Ethiopia Commodity Exchange introduced in late 2008 replaced the Auction System which was the internal system of selling coffee.

The Ethiopian Commodities Exchange also deals with other commodities beside coffee. All coffees sold through the Ethiopian Commodities Exchange are graded using an internal grading system from grade one to nine and allocated a sub-regional geographical designation where the coffee was produced and how it was processed.

Grading is based on physical appearance, number of defects and cup quality with grade one being the best while grade nine is the lowest quality. An example is Sidama C grade 3 where Sidama refers to the larger region where the coffee was grown while letter C refers to subregion.

All Ethiopian coffees are grouped into 4 groups, washed commercial, unwashed commercial, washed speciality and unwashed speciality.

Coffee grades 1 to 2 are considered as a speciality category while grades 3 to 9 are considered as commercial. Speciality grade coffees both washed and unwashed are graded as either Q1 or Q2.

As a general requirement coffee should possess a moisture content less than 11.5 % with 85% retained on screen size 14. Washed and unwashed commercial coffees are graded according to an internal scoring system.

Ethiopian Coffee Grades

Grade	Total score values (Green bean Cup quality score)
Grade 1	91-100
Grade 2	81-90
Grade 3	71-80
Grade 4	63-70
Grade 5	58-62
Grade 6	50-57
Grade 7	40-49
Grade 8	31-39
Grade 9	20-30

Source: ECX Coffee Contracts 2018

The grading score for the green bean is determined by the number of defects present, shape of the bean, colour, and odour.

Ethiopian Coffee Exchange coffee contracts are traded in standard lot sizes of 20 bags for small holders and 60 bags for coffee exporters.

After grading coffee is stored in ECX collection warehouses across the country for safe custody and a warehouse receipt is issued.

Coffee biding at the exchange is based on the grade of the coffee and trading is done through members or authorized representatives. Payment and ownership of coffee traded is transferred within 48 hours.

The Ethiopian Commodity Exchange promotes integrity of coffee quality, quantity and settlement of contracts beside disseminating market information to farmers.

Exporters have no access to samples for cupping or visual evaluation before bidding for the coffee. Disputes between members and the exchange are resolved by the Exchange Arbitration Tribunal.

After buying coffee at the exchange, the coffee is taken to the Dry Mill for milling and bulking ready for export. After dry milling a 3-kilogram sample is drawn at the Central Liquoring Unit from every lot of 300 bags bound for export through Djibouti and graded again.

Grading at the Central Liquoring Unit (CLU) is based on green bean physical appearance, defect counts and cup.

All coffees for export must meet the minimum respective grade standard threshold both in cup and defect count to qualify for export. If not, coffee is reprocessed again, commonly referred to as re-milling.

Exportable grades are restricted to grades 1 to 5. Small lots of coffees with similar cup profiles are bulked together to make up a container load requirement for export.

Ethiopia allows direct trade to the speciality market allowing producers to sell coffee directly to overseas roasters.

Ethiopian Market coffee Value Chain

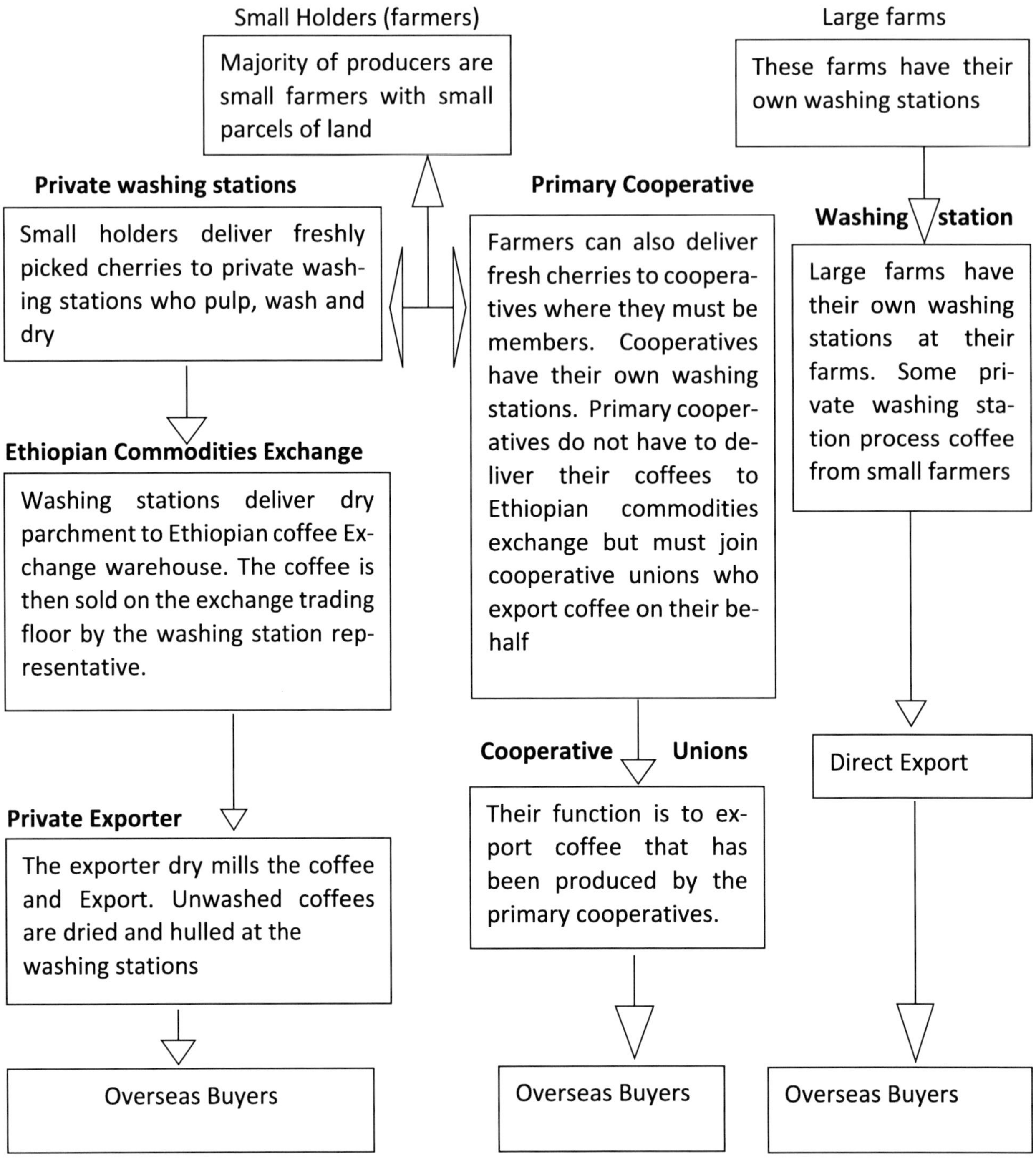

REFERENCES

1. Kenya Coffee Safari Brochure 1994
2. Kenya Gazette supplement no.51 (Legislative supplement no .28)
3. Coffee Board of Kenya Monthly bulletin 1994 April
4. Coffee Board of Kenya monthly bulletin July 1956
5. The coffee Act 2001 (No. 9 of 2001) 21st July 2006
6. Kenya Gazette supplement No. 96 (Acts No. 9): The Coffee General Amendment Rules.
7. The coffee quarterly issue no. 2/ 2006
8. Coffee grower's handbook (coffee research institute)
9. Coffee board of Kenya growers master file
10. Coffee- an exporters guide (commodity handbook)
11. Coffee board of Kenya circular No.6 dated 8th January 2006.
12. Kenya Gazette supplement no. 136 (5/10/2012)
13. Kenya Gazette supplement NO.100 1st July 2019, Crop regulations 2019.
14. Guidelines on implementation of direct sales of coffee
15. Coffee board of Kenya circular dated 28th July 1997
16. USDA Gain report NO.ET1904
17. Reuters September 9, 2010, article Kenya unveils new disease resistant coffee variety
18. Article by Danille Kilbride on Honey process (Perfect Daily Grind) on honey process February 23,2017.
19. International coffee organisation
20. Uganda Coffee Development Authority Manual (Arabica and Robusta Coffee Handbook)
21. Market and value chain survey (Uganda)EI-JR17053-NECMP-DB-VOL1-03
22. International trade centre 2016 (Coffee sector investment profile)
23. Circular on Tanzania coffee. (Kwa mujibu wa kanuni ya 63 kanuni za kahawa 2013)
24. ICO. document on national quality standards ICC 122-12, 23rd August 2018.
25. JICA Data collection Rwanda coffee growing and marketing. 2014
26. Uganda National coffee policy document 2013
27. ECX Coffee contracts 2010
28. Ethiopian coffee buying manual. Willem J Boot March 2011.
29. Tanzania Coffee Industry development strategy 2011-2021
30. USDA Tanzania annual coffee report 5-24-2018
31. Daily nation Sunday April 21, 2019.
32. Kenya ICO Coffee profile 124-7
33. Tanzania regulation 23(1)(2) of coffee regulation 2013
34. Coffee Flavour Chemistry by Ivon Flament et al, History of coffee).
35. (Article by Emmy Blotnick December 14, 2013.originally from Mental floss Magazine on five historical attempts to ban coffee)
36. Tanzania coffee board Declaration of the 2020/2021 Coffee buying season.
37. Where the wild coffee grows by Jeff Koehler 2017

38. History of East Africa by Keneth Higham 1963
39. Standard Newspaper June 8th, 2020
40. Article by Carley Clontz October 1,2018 Coffee versus the climate: Coffee production in Brazil
41. Kenya coffee bulletin August 1999

ACKNOWLEDGEMENTS

I would like to thank Stephen Kemp former Technical Services Manager of FreshFood Corporation Pty Ltd for his time and dedication in helping edit this book.

Special thanks to David Mbithi, CEO Nairobi Coffee Exchange for his support and for providing coffee industry data and historical photos.

Special thanks to the Chairman of the Thunguri Coffee Farmer's Co-operative Society for the valuable information and photos he has provided.

ABOUT THE AUTHOR

Sam Mungai is a qualified Q Grader who has worked in the coffee industry for more than 25 years. He started his career with the Coffee Board of Kenya where he trained and qualified as a coffee expert.

While working with the Coffee Board his roles included co-ordinating farmer's payments during the liberalization of the coffee industry, providing a link between the only miller then known as the Kenya Planters Cooperative Union and the Coffee Board, the sole marketing agent. He also contributed to the Coffee board of Kenya monthly Bulletin.

After leaving the Coffee Board he joined several coffee trading houses including Uneximp (Universal Import and Export) Ltd, C Dorman Ltd and Servicoff Ltd working as a quality expert. His role included quality control, auction buying and export operations.

After emigrating to Australia, he worked briefly for Micro Roast, a small coffee roaster before joining Freshfood Corporation Pty Ltd the second largest instant coffee manufacturer in Australia.

He is currently the Coffee Technologist responsible for the quality of imported green beans, roast and ground and instant coffee manufacture, new coffee blends and assisting in the development of value-added coffee products.

He is involved in Coffee Judging at the Royal Sydney show and MICE (Melbourne International Coffee Expo)

Made in United States
North Haven, CT
04 November 2026

21726210R00075